γ Gamma

Focus: Multiplication

Instruction Manual

by Steven P. Demme

Math·U·See

1-888-854-MATH (6284)

www.MathUSee.com

Math·U·See

1-888-854-MATH (6284)
www.MathUSee.com
Copyright © 2009 by Steven P. Demme

Graphic design by Christine Minnnich
Illustrations by Gregory Snader

Printed in the United States of America

γ Gamma

SCOPE & SEQUENCE

Math-U-See is a complete and comprehensive K-12 math curriculum. While each book focuses on a specific theme, Math-U-See continuously reviews and integrates topics and concepts presented in previous levels.

Primer

α Alpha | Focus: Single-Digit Addition and Subtraction

β Beta | Focus: Multiple-Digit Addition and Subtraction

γ Gamma | Focus: Multiplication

δ Delta | Focus: Division

ε Epsilon | Focus: Fractions

ζ Zeta | Focus: Decimals and Percents

Pre-Algebra

Algebra 1

Stewardship*

Geometry

Algebra 2

Pre Calculus with Trigonometry

Stewardship is a biblical approach to personal finance. The requisite knowledge for this curriculum is a mastery of the four basic operations, as well as fractions, decimals, and percents. In the Math-U-See sequence these topics are thoroughly covered in Alpha through Zeta. We also recommend Pre-Algebra and Algebra 1 since over half of the lessons require some knowledge of algebra. Stewardship may be studied as a one-year math course or in conjunction with any of the secondary math levels.

Five Minutes for Success

Welcome to *Gamma*. I believe you will have a positive experience with the unique Math-U-See approach to teaching math. These first few pages explain the essence of this methodology, which has worked for thousands of students and teachers. I hope you will take five minutes and read through these steps carefully.

I am assuming your student has a thorough grasp of addition and subtraction.

If you are using the program properly and still need additional help, you may contact your authorized representative, or visit Math-U-See online at http://www.mathusee.com/support.html

— S. Demme

The Goal of Math-U-See

The underlying assumption or premise of Math-U-See is that the reason we study math is to apply math in everyday situations. Our goal is to help produce confident problem solvers who enjoy the study of math. These are students who learn their math facts, rules, and formulas and are able to use this knowledge to solve word problems and real life applications. Therefore, the study of math is much more than simply committing to memory a list of facts. It includes memorization, but it also encompasses learning the underlying concepts of math that are critical to successful problem solving.

More than Memorization

Many people confuse memorization with understanding. Once while I was teaching seven junior high students, I asked how many pieces they would each receive if there were fourteen pieces. The students' response was, "What do we do: add, subtract, multiply, or divide?" Knowing how to divide is important, understanding when to divide is equally important.

THE SUGGESTED 4-STEP MATH-U-SEE APPROACH

In order to train students to be confident problem solvers, here are the four steps that I suggest you use to get the most from the Math-U-See curriculum.

Step 1. Prepare for the Lesson
Step 2. Present the New Topic
Step 3. Practice for Mastery
Step 4. Progression after Mastery

Step 1. Prepare for the Lesson.

Watch the DVD to learn the new concept and see how to demonstrate this concept with the manipulatives when applicable. Study the written explanations and examples in the instruction manual. Many students watch the DVD along with their instructor.

Step 2. Present the New Topic

Now that you have studied the new topic, choose problems from the first lesson practice page to present the new concept to your students.

a. Build: Use the manipulatives to demonstrate the problems from the worksheet.

b. Write: Record the step-by-step solutions on paper as you work them through with the manipulatives.

c. Say: Explain the *why* and *what* of math as you build and write.

Do as many problems as you feel are necessary until the student is comfortable with the new material. One of the joys of teaching is hearing a student say, *"Now I get it!"* or *"Now I see it!"*

Step 3. Practice for Mastery.

Using the examples and the lesson practice problems from the student text, have the students practice the new concept until they understand it. It is one thing for students to watch someone else do a problem, it is quite another to do the same problem themselves. Do enough examples together until they can do them without assistance.

Do as many of the lesson practice pages as necessary (not all pages may be needed) until the students remember the new material and gain understanding. Give special attention to the word problems, which are designed to apply the concept being taught in the lesson.

Another resource is the Math-U-See web site, which has online drill and downloadable worksheets for more practice. Go to www.mathusee.com and select "Online Helps."

Step 4. Progression after Mastery.

Once mastery of the new concept is demonstrated, proceed to the systematic review pages for that lesson. Mastery can be demonstrated by having each student teach the new material back to you. The goal is not to fill in worksheets, but to be able to teach back what has been learned.

The systematic review worksheets review the new material, as well as provide practice of the math concepts previously studied. Remediate missed problems as they arise to ensure continued mastery.

Proceed to the lesson tests. These were designed to be an assessment tool to help determine mastery, but they may also be used as extra worksheets. Your students will be ready for the next lesson only after demonstrating mastery of the new concept and continued mastery of concepts found in the systematic review worksheets.

Confucius was reputed to have said, "Tell me, I forget; Show me, I understand; Let me do it, I will remember." To which we add, **"Let me teach it and I will have achieved mastery!"**

Length of a Lesson

So how long should a lesson take? This will vary from student to student and from topic to topic. You may spend a day on a new topic, or you may spend several days. There are so many factors that influence this process that it is impossible to predict the length of time from one lesson to another. I have spent three days on a lesson and I have also invested three weeks in a lesson. This occurred in the same book with the same student. If you move from lesson to lesson too quickly without the student demonstrating mastery, he will become overwhelmed and discouraged as he is exposed to more new material without having learned the previous topics. But if you move too slowly, your student may become bored and lose interest in math. I believe that as you regularly spend time working along with your student, you will sense when is the right time to take the lesson test and progress through the book.

By following the four steps outlined above, you will have a much greater opportunity to succeed. Math must be taught sequentially, as it builds line upon line and precept upon precept on previously learned material. I hope you will try this methodology and move at your student's pace. As you do, I think you will be helping to create a confident problem solver who enjoys the study of math.

ONGOING SUPPORT
AND ADDITIONAL RESOURCES

Welcome to the Math-U-See Family!

Now that you have invested in your children's education, I would like to tell you about the resources that are available to you. Allow me to introduce you to your regional representative, our ever improving website, the Math-U-See blog, our new free e-mail newsletter, the online Forum, and the Users Group.

Most of our regional **Representatives** have been with us for over 10 years. What makes them unique is their desire to serve and their expertise. They have all used Math-U-See and are able to answer most of your questions, place your student(s) in the appropriate level, and provide knowledgeable support throughout the school year. They are wonderful!

Come to your local curriculum fair where you can meet your rep face-to-face, see the latest products, attend a workshop, meet other MUS users at the booth, and be refreshed. We are at most curriculum fairs and events. To find the fair nearest you, click on "Events Calendar" under "News."

The **Website**, at www.mathusee.com, is continually being updated and improved. It has many excellent tools to enhance your teaching and provide more practice for your student(s).

 ## ONLINE DRILL

Let your students review their math facts online. Just enter the facts you want to learn and start drilling. This is a great way to commit those facts to memory.

 ## WORKSHEET GENERATOR

Create custom worksheets to print out and use with your students. It's easy to use and gives you the flexibility to focus on a specific lesson. Best of all — it's free!

Math-U-See Blog

Interesting insights and up-to-date information appear regularly on the Math-U-See Blog. The blog features updates, rep highlights, fun pictures, and stories from other users. Visit us and get the latest scoop on what is happening .

Email Newsletter

For the latest news and practical teaching tips, sign up online for the free Math-U-See e-mail newsletter. Each month you will receive an e-mail with a teaching tip from Steve as well as the latest news from the website. It's short, beneficial, and fun. Sign up today!

The Math-U-See Forum and the Users Group put the combined wisdom of several thousand of your peers with years of teaching experience at your disposal.

Online Forum

Have a question, a great idea, or just want to chitchat with other Math-U-See users? Go to the online forum. You can also use the forum to post a specific math question if you are having difficulty in a certain lesson. Head on over to the forum and join in the discussion.

Yahoo Users Group

The MUS-users group was started in 1998 for lovers and users of the Math-U-See program. It was founded by two home-educating mothers and users of Math-U-See. The backbone of information and support is provided by several thousand fellow MUS users.

For Specific Math Help

When you have watched the DVD instruction and read the instruction manual and still have a question, we are here to help. Call your local rep, click the support link and e-mail us here at the home office, or post your question on the forum. Our trained staff have used Math-U-See themselves and are available to answer a question or walk you through a specific lesson.

Feedback

Send us an e-mail by clicking the feedback link. We are here to serve you and help you teach math. Ask a question, leave a comment, or tell us how you and your student are doing with Math-U-See.

Our hope and prayer is that you and your students will be equipped to have a successful experience with math!

Blessings,

Steve Demme

Rectangles, Factors and Products

Rectangle means right angle. "Rect" comes from a German word that means right. A *right angle* is a square corner. If you find an object with four square corners, it is a rectangle. Look around you and see how many rectangles there are. This piece of paper is a rectangle. How many others can you identify?

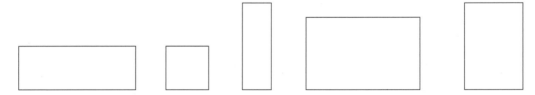

A *square* is a special kind of rectangle. It has four right angles, so it is a rectangle. But it also has four sides that are the same length, so it is a square.

A rectangle is measured by its dimensions. A *dimension* is the length of a side. In the following picture, how long is the over dimension, and how long is the up dimension?

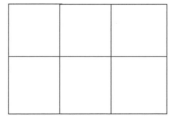

The over dimension is three and the up dimension is two. The rectangle also has area. You see that by the squares inside it. The dimensions tell how long the sides or edges are. The area tells how many squares are in the inside. In this rectangle, the area is six.

What is the over dimension and the up dimension? What is the area?

The over dimension is 5
and the up dimension is 3.
The area is 15 square units.

Later we will use the words *factor* instead of dimension and **product** instead of area. By learning the skip-counting facts and how to build rectangles, we are laying a foundation for multiplication.

On the worksheets there are rectangles that look like the figures below. Write the dimensions of each rectangle in the parentheses and the area in the oval.

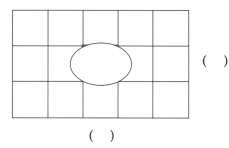

Example 1

Build a rectangle with dimensions over 10 and up 5. We can write this as
(10)(5). Now find the area.
→ ↑

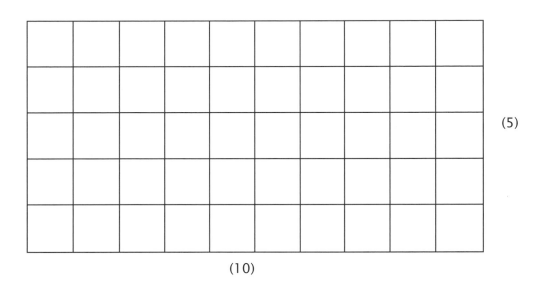

(5)

(10)

Skip counting by 5 (5-10-15-20-25-30) or by 10 (10-20-30-40-50),
we find an area of 50.

Another way to think of this is to put a piece of paper over the rectangle in order to emphasize that the dimensions are 10 and 5 and that these are just how long the sides are. Then lift off the paper, and you can see that the area is the product.

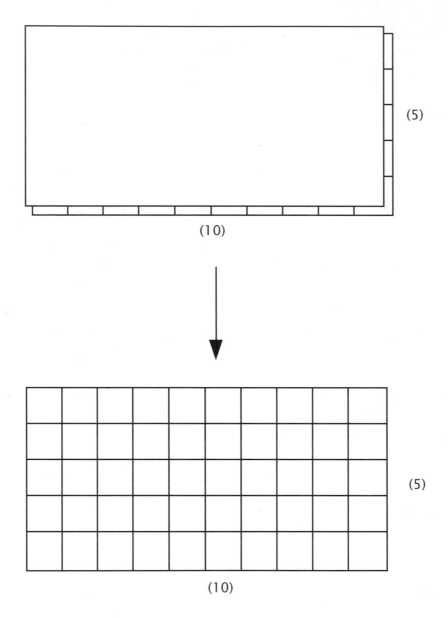

Multiply by 1 and 0, Commutative Property

There is an interactive math facts practice page available online at mathusee.com/drillsheet.html.

Multiplication is fast adding of the same number. Tell the student that one multiplied by three is simply one counted three times or 1 + 1 + 1. The facts in this lesson are the easiest facts to learn. But while the facts are simple, the symbolism is new. There are three ways to illustrate multiplication: "x" (the most common), the floating dot " • ," and two parentheses side by side ( )( ). So far, the student has learned one symbol for addition "+" and one for subtraction "-," so comment that there are three symbols for fast adding or multiplication.

You may verbalize multiplication several ways. The problem 1 x 3 could be "one counted three times," "one times three," or "one multiplied by three." Multiplication is fast adding of the same number, so it can also be shown as 1 + 1 + 1 = 3.

In the example, 1 x 3 is illustrated with a rectangle, and the three ways of writing it are given.

Example 1

1 x 3 = 3

1 • 3 = 3

(1)(3) = 3

At some point, a student will ask whether 1 x 3 is the same as 3 x 1. This is a golden opportunity to teach that multiplication is commutative. That is, you can change the order of the factors without changing the product. To show this, simply build the rectangle vertically and horizontally to show that both are the same rectangle, except that one is standing and one is lying down. ☺ Just as in addition, the problems themselves may be written vertically and horizontally.

Example 2

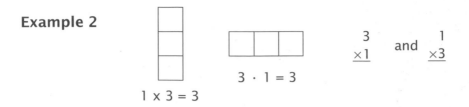

$$1 \times 3 = 3$$

$$3 \cdot 1 = 3$$

$$\begin{array}{r} 3 \\ \times 1 \\ \hline \end{array} \quad \text{and} \quad \begin{array}{r} 1 \\ \times 3 \\ \hline \end{array}$$

Multiplication by zero is also easier than you may think. When you verbalize it, mention the zero factor first, as this makes it much easier to grasp. The problem 0 x 5 is zero counted five times or 0 + 0 + 0 + 0 + 0, which is obviously 0. I like to use a word problem to make this clear. I say, "What if you didn't earn any money today or for the next four days, how much would you have?" "0 + 0 + 0 + 0 + 0, or zero counted five days, or zero money." Use something similar to illustrate this concept, since we obviously can't show zero with manipulatives!

This is a good time to show all of the multiplication facts on a chart. After the student learns several facts, color or circle the ones learned to encourage him in his progress. I will put a small chart in each of the lessons in the instruction manual and circle the facts currently being studied, as well as those already learned. There is a chart for the student after the lesson 2 worksheets in the student text.

0 x 0	0 x 1	0 x 2	0 x 3	0 x 4	0 x 5	0 x 6	0 x 7	0 x 8	0 x 9	0 x 10
1 x 0	1 x 1	1 x 2	1 x 3	1 x 4	1 x 5	1 x 6	1 x 7	1 x 8	1 x 9	1 x 10
2 x 0	2 x 1	2 x 2	2 x 3	2 x 4	2 x 5	2 x 6	2 x 7	2 x 8	2 x 9	2 x 10
3 x 0	3 x 1	3 x 2	3 x 3	3 x 4	3 x 5	3 x 6	3 x 7	3 x 8	3 x 9	3 x 10
4 x 0	4 x 1	4 x 2	4 x 3	4 x 4	4 x 5	4 x 6	4 x 7	4 x 8	4 x 9	4 x 10
5 x 0	5 x 1	5 x 2	5 x 3	5 x 4	5 x 5	5 x 6	5 x 7	5 x 8	5 x 9	5 x 10
6 x 0	6 x 1	6 x 2	6 x 3	6 x 4	6 x 5	6 x 6	6 x 7	6 x 8	6 x 9	6 x 10
7 x 0	7 x 1	7 x 2	7 x 3	7 x 4	7 x 5	7 x 6	7 x 7	7 x 8	7 x 9	7 x 10
8 x 0	8 x 1	8 x 2	8 x 3	8 x 4	8 x 5	8 x 6	8 x 7	8 x 8	8 x 9	8 x 10
9 x 0	9 x 1	9 x 2	9 x 3	9 x 4	9 x 5	9 x 6	9 x 7	9 x 8	9 x 9	9 x 10
10 x 0	10 x 1	10 x 2	10 x 3	10 x 4	10 x 5	10 x 6	10 x 7	10 x 8	10 x 9	10 x 10

WORD PROBLEM TIPS

Parents often find it challenging to teach children how to solve word problems. Here are some suggestions for helping your student learn this important skill.

The first step is to realize that word problems require both reading and math comprehension. Don't expect a child to be able to solve a word problem if he does not thoroughly understand the math concepts involved. On the other hand, a student may have a math skill level that is stronger than his or her reading- comprehension skills. Below are a number of strategies to improve comprehension skills in the context of story problems. You may decide which ones work best for you and your child.

Strategies for word problems:

1. Ignore numbers at first and read the story. It may help some students to read the question aloud. Every word problem tells a story. Before deciding what math operation is required, let the student retell the story in his own words. Who is involved? Are they receiving gifts, losing something, or dividing a treat?

2. Relate the story to real life, perhaps by using names of family members. For some students, this makes the problem more interesting and relevant.

3. Build, draw, or act out the story. Use the blocks or actual objects when practical. Especially in the lower levels, you may require the student to use the blocks for word problems, even when the facts have been learned. Don't be afraid to use a little drama as well. The purpose is to make it as real and meaningful as possible.

4. Look for the common language used in a particular kind of problem. Pay close attention to the word problems on the lesson practice pages, as they model different kinds of language that may be used for the new concept just studied. For example, "altogether" indicates addition. These "key words" can be useful clues but should not be a substitute for understanding.

5. Look for practical applications that use the concept and ask questions in that context.

6. Have the student invent word problems to illustrate the number problems from the lesson.

Cautions:

1. Unneeded information may be included in the problem. For example, we may be told that Suzie is eight years old, but the eight is irrelevant when adding up the number of gifts she received.

2. Some problems may require more than one step to solve. Model these questions carefully.

3. There may be more than one way to solve some problems. Experience will help the student choose the easier or preferred method.

4. Estimation is a valuable tool for checking an answer. If an answer is unreasonable, it is possible that the wrong method was used to solve the problem.

Multiplication Facts Sheet

0×0	0×1	0×2	0×3	0×4	0×5	0×6	0×7	0×8	0×9	0×10
1×0	1×1	1×2	1×3	1×4	1×5	1×6	1×7	1×8	1×9	1×10
2×0	2×1	2×2	2×3	2×4	2×5	2×6	2×7	2×8	2×9	2×10
3×0	3×1	3×2	3×3	3×4	3×5	3×6	3×7	3×8	3×9	3×10
4×0	4×1	4×2	4×3	4×4	4×5	4×6	4×7	4×8	4×9	4×10
5×0	5×1	5×2	5×3	5×4	5×5	5×6	5×7	5×8	5×9	5×10
6×0	6×1	6×2	6×3	6×4	6×5	6×6	6×7	6×8	6×9	6×10
7×0	7×1	7×2	7×3	7×4	7×5	7×6	7×7	7×8	7×9	7×10
8×0	8×1	8×2	8×3	8×4	8×5	8×6	8×7	8×8	8×9	8×10
9×0	9×1	9×2	9×3	9×4	9×5	9×6	9×7	9×8	9×9	9×10
10×0	10×1	10×2	10×3	10×4	10×5	10×6	10×7	10×8	10×9	10×10

Skip Count by 2, 5, and 10

Skip counting is the ability to count groups of the same number quickly. For example, if you were to skip count by three you would skip the one and the two and say "three," skip the four and the five and say "six," and follow with "9–12–15–18," etc. Skip counting by seven is 7–14–21–28–35–42–49–56–63–70.

Here are five reasons for learning skip counting:

1. This skill lays a solid foundation for learning your multiplication facts. The problem 3 + 3 + 3 + 3 can be written as 3 x 4. If a child can skip count, he could say "3–6–9–12." Then he could read 3 x 4 as "three counted four times is twelve." As you learn your skip-counting facts, you are learning all of the products of the multiplication facts in order. Multiplication is fast adding of the same number, and skip counting illustrates this beautifully. You can think of multiplication as a shortcut to the skip counting process. Consider 3 x 5. I could skip count by three five times (3–6–9–12–15) to come up with the solution. Or, after I learn my facts, I can say, "Three counted five times is fifteen." The latter is much faster.

2. Skip counting teaches the concept of multiplication. I had a teacher tell me that her students had successfully memorized their facts but didn't know what they had acquired. After learning the skip-counting facts, they understood what they had learned. I used to say that multiplying was fast adding. In reality, it is fast adding of the same number. I can't multiply to find the solution to 1 + 4 + 6 + 9, but I can multiply to solve 4 + 4 + 4. Skip counting reinforces and teaches the concept of multiplication.

3. As a skill in itself, multiple counting is helpful. A pharmacist attending a workshop told me he skip counted when counting pills as they went into the bottles. Another man said he used the same skill for counting inventory at the end of every workday.

4. It teaches you the multiples of a number, which are so important when making equivalent fractions and finding common denominators. 2/5 = 4/10 = 6/15 = 8/20. The numbers 2–4–6–8 are multiples of 2, and 5–10–15–20 are multiples of 5.

5. In the curriculum, skip counting is reviewed in sequence form, asking students to fill in the blanks: __, __, __, 12, 15, __, __, __, 27, 30. This encourages them to find patterns in math, and patterns are key to understanding this logical and important subject.

One way to learn the skip-counting facts is with the *Skip Count and Addition Songbook.* Included is a CD with the skip count songs from the twos to the nines sung to tunes taken from hymns and Christmas carols. Children enjoy it, and it has proven very effective.

Another way to teach skip counting is by just plain counting and then beginning to skip some of the numbers. This is where the concept originated. Begin by counting each square: 1–2–3–4–5–6 . . . through 20. After this sequence is learned, skip the first number and just count the second: 2–4–6 . . . through 20. This is skip counting. You know it as counting by two.

When first introducing this, you might try pointing to each square, and as you count the first number quietly, ask the student(s) to say the second number loudly. Continue this practice, doing it quieter each time, until you are just pointing to the first block silently while encouraging them to say the number loudly when you point to the second square. The student(s) see it, hear it, and say it, and you can write the facts 2–4–6 . . . 20 as well.

Example 1

Skip count and write the numbers on the lines. Say each number out loud as you count and write. After filling out the rectangle, write the numbers in the spaces provided beneath the figure. By skipping the first space, we come up with the expression "skip counting." The solution follows at the bottom of the page.

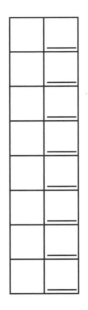

__2__ , __4__ , _____ , _____ , _____ , _____ , _____ , _____

Solution

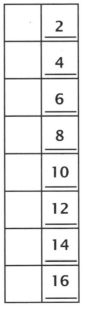

__2__ , __4__ , __6__ , __8__ , __10__ , __12__ , __14__ , __16__

Skip Count by 5 - Also in this lesson, we're reviewing skip counting by five. Use the same techniques to introduce and teach this important skill as you did for the twos. Some practical examples are fingers on one hand, toes on a foot, pennies in a nickel, players on a basketball team, and sides of a pentagon. We can also remind the student that one nickel has the same value as five pennies, and apply the skill of counting by five to find out how many pennies in several nickels.

Example 2

As with the twos, skip count and write the numbers on the lines. Say each number out loud as you count and write. After filling out the rectangle, write the numbers in the spaces provided beneath the figure.

				5
				10
				15
				20

__5__, ____, ____, ____,

Notice that skip counting by five, or fast counting by five, is a way of illustrating multiplication. The rectangle above shows five counted four times or 5 x 4, which is 20. When students learn all of these skip counting facts, they have learned all of the answers to the five facts. Please don't proceed to the next lesson until the student can skip count by five to 50.

Example 3

Fill in the missing information on the lines.

__5__, ____, __15__, ____, ____, __30__, ____, __40__, __45__, ____

Solution

__5__, __10__, __15__, __20__, __25__, __30__, __35__, __40__, __45__, __50__

Skip Count by 10 - After you learn the twos and fives, review skip counting by ten. Some practical examples are fingers on both hands, toes on both feet, and pennies in a dime.

Multiply by 2, 1 Quart = 2 Pints

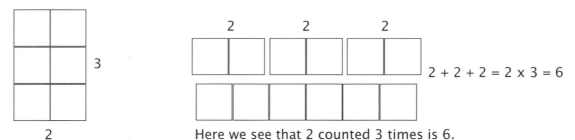

3

2

The dimensions, or factors, are 2 and 3, and the area, or product, is 6.

2 + 2 + 2 = 2 x 3 = 6

Here we see that 2 counted 3 times is 6.

A rectangle has two components, the **dimensions** and the **area**. The dimensions of this rectangle are two and three. Some say the base is two and the height is three. Others say the length is three and the width is two. Either way is accurate, but for our purposes, we will say the over dimension is two and the up dimension is three.

Besides the dimensions, there is also the area. In this rectangle, the area would be six square units. Instead of dimensions, we are going to call the over and the up the *factors*. Instead of area we will use the word ***product***. The problem 2 x 3 = 6 is illustrated with our rectangle above. This may be verbalized as "two counted three times," "two times three," or "two multiplied by three."

Remember that because of the commutative property, we can also describe the rectangle as 3 x 2 = 6.

Now that the student can skip count by two, we begin memorizing the specific two facts.

One way to show the two facts is a progression, with the blocks increasing by two each time. If you use the colored unit bars to represent the 2, 4, 6, and 8 bars, a pattern emerges that is reinforced on the chart below. Notice this as you move from right to left: orange (2)–yellow (4)–violet (6)–brown (8). The pattern repeats beginning with 12.

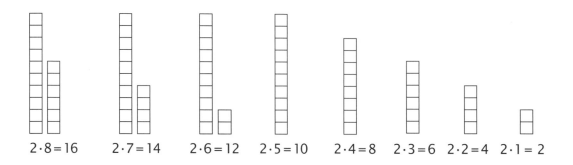

$2 \cdot 8 = 16$ $2 \cdot 7 = 14$ $2 \cdot 6 = 12$ $2 \cdot 5 = 10$ $2 \cdot 4 = 8$ $2 \cdot 3 = 6$ $2 \cdot 2 = 4$ $2 \cdot 1 = 2$

Another way to show the two facts is on a number chart. Circling all of the answers to the two facts, or multiples of two, reveals an interesting pattern that corresponds with the blocks above.

1 ② 3 ④ 5 ⑥ 7 ⑧ 9 ⑩
11 ⑫ 13 ⑭ 15 ⑯ 17 ⑱ 19 ⑳

Since the student has learned the skip counting facts and is able to say them without music, he or she knows all of the multiples or products of two. This is a good time for you as the teacher to put the factors with the products. Say the factors slowly, then ask the student to say the product, which is the skip counting. For example, you say "two counted one time" or "two times one," and the student says "two." You continue by saying "two times two," and the student says "four."

Here are the two facts with the corresponding numbering.

$$\frac{0}{2\times0} \quad \frac{2}{2\times1} \quad \frac{4}{2\times2} \quad \frac{6}{2\times3} \quad \frac{8}{2\times4} \quad \frac{10}{2\times5} \quad \frac{12}{2\times6} \quad \frac{14}{2\times7} \quad \frac{16}{2\times8} \quad \frac{18}{2\times9} \quad \frac{20}{2\times10}$$

       ↑   ↑   ↑

2 counted 2 times   2 counted 5 times   2 counted 9 times

0 x 0	0 x 1	0 x 2	0 x 3	0 x 4	0 x 5	0 x 6	0 x 7	0 x 8	0 x 9	0 x 10
1 x 0	1 x 1	1 x 2	1 x 3	1 x 4	1 x 5	1 x 6	1 x 7	1 x 8	1 x 9	1 x 10
2 x 0	2 x 1	2 x 2	2 x 3	2 x 4	2 x 5	2 x 6	2 x 7	2 x 8	2 x 9	2 x 10
3 x 0	3 x 1	3 x 2	3 x 3	3 x 4	3 x 5	3 x 6	3 x 7	3 x 8	3 x 9	3 x 10
4 x 0	4 x 1	4 x 2	4 x 3	4 x 4	4 x 5	4 x 6	4 x 7	4 x 8	4 x 9	4 x 10
5 x 0	5 x 1	5 x 2	5 x 3	5 x 4	5 x 5	5 x 6	5 x 7	5 x 8	5 x 9	5 x 10
6 x 0	6 x 1	6 x 2	6 x 3	6 x 4	6 x 5	6 x 6	6 x 7	6 x 8	6 x 9	6 x 10
7 x 0	7 x 1	7 x 2	7 x 3	7 x 4	7 x 5	7 x 6	7 x 7	7 x 8	7 x 9	7 x 10
8 x 0	8 x 1	8 x 2	8 x 3	8 x 4	8 x 5	8 x 6	8 x 7	8 x 8	8 x 9	8 x 10
9 x 0	9 x 1	9 x 2	9 x 3	9 x 4	9 x 5	9 x 6	9 x 7	9 x 8	9 x 9	9 x 10
10 x 0	10 x 1	10 x 2	10 x 3	10 x 4	10 x 5	10 x 6	10 x 7	10 x 8	10 x 9	10 x 10

1 Quart = 2 Pints - After the student has learned the two facts to your satisfaction, measurement is a good place to apply this newly acquired skill. Begin with two one-pint containers and one one-quart container. Fill up the one-quart container and empty it into the two one-pint containers. The student can see that one quart equals two pints. Or do the converse by emptying the two pints into the one quart. Then use multiplying by two to find out how many pints are in the quarts.

Example 1

How many pints are in three quarts?

 3 quarts x 2 (pints in one quart) = 6 pints
There are six pints in three quarts.

Example 2

How many pints are in the quarts that are shown?

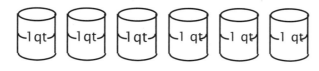 6 x 2 = 12

Multiply by 10, 10¢ = 1 Dime

When multiplying by 10, encourage the student to look for patterns. Notice that whenever you multiply 10 times any number, the answer is that number plus a zero. That is because 10 is made up of a "1" digit and a "0" digit. So 4 x 10 is 4 x 1 = 4 *and* 4 x 0 = 0, or 40. To make sure the student has this concept, I like to ask, "What is banana times 10?" The answer is "banana zero" pronounced "banana-ty." The "ty" stands for 10. These are easy facts to learn and remember, but don't take them for granted. Make sure they are mastered using any of the techniques shown below.

On the worksheets, there have been rectangles where the student wrote in the fact at the end of the line in the space with an underline. These can be put to another use by adding the multiplication problem to the multiple of 10. Here are a few examples.

[_____ 10]	Ten counted one time equals ten or 10 x 1 = 10.
[_____ 20]	Ten counted two times equals twenty or 10 x 2 = 20.
[_____ 30]	Ten counted three times equals thirty or 10 x 3 = 30.
[_____ 40]	Ten counted four times equals forty or 10 x 4 = 40.

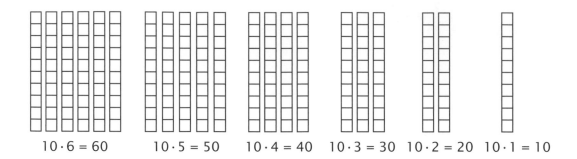

10·6 = 60   10·5 = 50   10·4 = 40   10·3 = 30   10·2 = 20   10·1 = 10

Another way to show this is on a number chart. Circling all of the 10 facts, or multiples of 10, reveals the pattern that corresponds to the blocks above.

⓪	1	2	3	4	5	6	7	8	9
⑩	11	12	13	14	15	16	17	18	19
⑳	21	22	23	24	25	26	27	28	29
㉚	31	32	33	34	35	36	37	38	39
㊵	41	42	43	44	45	46	47	48	49
㊿	51	52	53	54	55	56	57	58	59
60	61	62	63	64	65	66	67	68	69
70	71	72	73	74	75	76	77	78	79
80	81	82	83	84	85	86	87	88	89
90	91	92	93	94	95	96	97	98	99
100									

Of course each fact can be built in the shape of a rectangle. Whenever illustrating with the blocks, also write it and say it as you build.

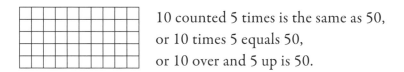

10 counted 5 times is the same as 50,
or 10 times 5 equals 50,
or 10 over and 5 up is 50.

Counting by 10 is the first step. After this is accomplished, say the factors slowly, and then ask the student to say the product. For example, you say "ten counted one time," or "10 times one," and the student says "10." Continue by saying "10 times 2," and the student says "20." (I often have the student say "two-ty" as well as 20 to show there is order in our words.) Proceed through all the facts sequentially just as when the student learned to count by 10.

Here are the 10 facts with the corresponding numbering.

0	10	20	30	40	50	60	70	80	90	100
(10)(0)	(10)(1)	(10)(2)	(10)(3)	(10)(4)	(10)(5)	(10)(6)	(10)(7)	(10)(8)	(10)(9)	(10)(10)

↑   10 counted 1 time   ↑   10 counted 4 times   ↑   10 counted 9 times

0 x 0	0 x 1	0 x 2	0 x 3	0 x 4	0 x 5	0 x 6	0 x 7	0 x 8	0 x 9	0 x 10
1 x 0	1 x 1	1 x 2	1 x 3	1 x 4	1 x 5	1 x 6	1 x 7	1 x 8	1 x 9	1 x 10
2 x 0	2 x 1	2 x 2	2 x 3	2 x 4	2 x 5	2 x 6	2 x 7	2 x 8	2 x 9	2 x 10
3 x 0	3 x 1	3 x 2	3 x 3	3 x 4	3 x 5	3 x 6	3 x 7	3 x 8	3 x 9	3 x 10
4 x 0	4 x 1	4 x 2	4 x 3	4 x 4	4 x 5	4 x 6	4 x 7	4 x 8	4 x 9	4 x 10
5 x 0	5 x 1	5 x 2	5 x 3	5 x 4	5 x 5	5 x 6	5 x 7	5 x 8	5 x 9	5 x 10
6 x 0	6 x 1	6 x 2	6 x 3	6 x 4	6 x 5	6 x 6	6 x 7	6 x 8	6 x 9	6 x 10
7 x 0	7 x 1	7 x 2	7 x 3	7 x 4	7 x 5	7 x 6	7 x 7	7 x 8	7 x 9	7 x 10
8 x 0	8 x 1	8 x 2	8 x 3	8 x 4	8 x 5	8 x 6	8 x 7	8 x 8	8 x 9	8 x 10
9 x 0	9 x 1	9 x 2	9 x 3	9 x 4	9 x 5	9 x 6	9 x 7	9 x 8	9 x 9	9 x 10
10 x 0	10 x 1	10 x 2	10 x 3	10 x 4	10 x 5	10 x 6	10 x 7	10 x 8	10 x 9	10 x 10	←

↑

10¢ = 1 Dime - A good place to apply math is with money. We've learned that 10¢ is the same as one dime, so we can ask how many pennies in six dimes to apply 6 x 10. The answer is 60¢.

Dime   10¢   1¢   1¢   1¢   1¢   1¢   1¢   1¢   1¢   1¢   1¢

Example

How many pennies, or cents, in six dimes?

10¢   10¢   10¢   10¢   10¢   10¢

6 · 10¢ = 60¢

We will be reviewing and using multiplication facts throughout the student textbook. If you find that you need more review of the multiplication facts, consult the Math-U-See Web site, which provides online drill and downloadable work sheets. Go to www.mathusee.com and click on E-Sources.

LESSON 6

Multiply by 5, 5¢ = 1 Nickel

After the two facts and ten facts have been mastered, we turn our attention to the five facts. This is a good time to teach odd and even numbers. Evens are multiples of two and end in 0, 2, 4, 6, and 8. Odd numbers end in 1, 3, 5, 7, and 9. An even number times five will end in zero. An odd number times five will end in five. Notice the pattern that emerges and reinforce this using the manipulatives.

Multiplying 5 by 1, 3, 5, and 7 yields products that end in 5 (5, 15, 25, and 35). Multiplying 5 by 2, 4, and 6 gives answers that end in 0 (10, 20, and 30).

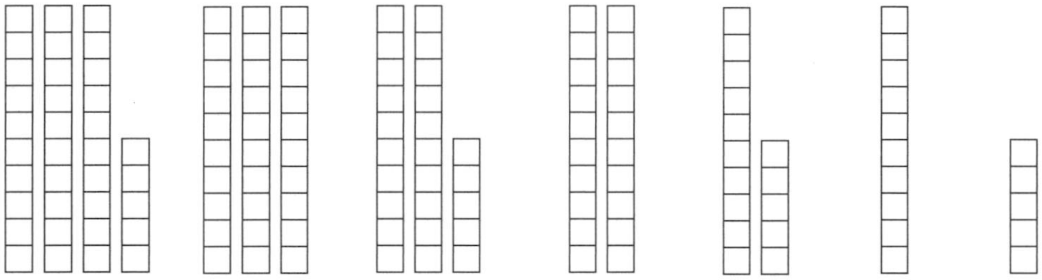

$5 \cdot 7 = 35$ $5 \cdot 6 = 30$ $5 \cdot 5 = 25$ $5 \cdot 4 = 20$ $5 \cdot 3 = 15$ $5 \cdot 2 = 10$ $5 \cdot 1 = 5$

Another way to show this is on a number chart. Circling all of the five facts, or multiples of five, reveals an interesting pattern that corresponds to the block pattern shown above.

⓪	1	2	3	4	⑤	6	7	8	9
⑩	11	12	13	14	⑮	16	17	18	19
⑳	21	22	23	24	㉕	26	27	28	29
㉚	31	32	33	34	㉟	36	37	38	39
㊵	41	42	43	44	㊺	46	47	48	49
㊿	51	52	53	54	(55)	56	57	58	59

Since the student has learned the skip-counting facts and is able to say them without music, he knows all of the multiples of five, or products of five. This is a good time for you as the teacher to put the factors with the product. Say the factors slowly, and then ask the student to say the product, which is skip counting. For example, you say "five counted one time" or "five times one" and the student says "five." You continue by saying "five times two" and the student says "ten." Proceed through all the facts sequentially just as they learned the skip-counting facts.

Here are the five facts with the corresponding numbering.

0	5	10	15	20	25	30	35	40	45	50
5×0	5×1	5×2	5×3	5×4	5×5	5×6	5×7	5×8	5×9	5×10

↑   ↑   ↑

"5 counted 1 time"   "5 counted 6 times"   "5 counted 10 times"

Of course, each fact can be built in the shape of a rectangle. Whenever illustrating with the blocks, also write it and say it as you build.

5 counted five times is the same as 25,
or 5 times 5 equals 25,
or 5 over and 5 up is 25.

On the worksheets, there have been rectangles where the student wrote in the fact at the end of the line, in the space with an underline. These can be put to another use by adding the multiplication problem to the multiple of five. Here are a few examples.

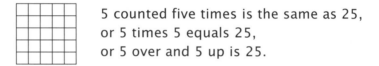

Five counted one time equals five or 5 x 1 = 5.
Five counted two times equals ten or 5 x 2 = 10.
Five counted three times equals fifteen or 5 x 3 = 15.

Another pattern I've observed that helps to learn the five facts builds on what we have already learned with the ten facts. There are two ways to do this. In the first approach, recall that 10 x 6 = 60, and half of that is 30. Because 5 is half of 10, 5 x 6 is 30. Here we first multiply by 10 and then cut the answer in half. You can build this to illustrate the concept. This presumes that the student knows how to take a half of something. The picture on the next page illustrates this.

In the second approach, instead of multiplying by 10 (adding a zero to the factor), and then cutting the answer in half, we cut the factor in half, and then multiply by 10. For 5 x 6, take half of 6, which is 3, and then multiply by 10 (add a zero). These approaches work for all even numbers.

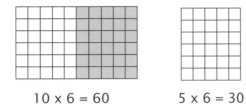

10 x 6 = 60   5 x 6 = 30   or   Six fives is the same as three tens.

Example 1

5 x 12   Take half of 12 and add a 0 to get 60.

                    or

5 x 12   Multiply 12 by 10 (add a 0) to get 120,
                and then cut the answer in half to get 6.

0 x 0	0 x 1	0 x 2	0 x 3	0 x 4	0 x 5	0 x 6	0 x 7	0 x 8	0 x 9	0 x 10
1 x 0	1 x 1	1 x 2	1 x 3	1 x 4	1 x 5	1 x 6	1 x 7	1 x 8	1 x 9	1 x 10
2 x 0	2 x 1	2 x 2	2 x 3	2 x 4	2 x 5	2 x 6	2 x 7	2 x 8	2 x 9	2 x 10
3 x 0	3 x 1	3 x 2	3 x 3	3 x 4	3 x 5	3 x 6	3 x 7	3 x 8	3 x 9	3 x 10
4 x 0	4 x 1	4 x 2	4 x 3	4 x 4	4 x 5	4 x 6	4 x 7	4 x 8	4 x 9	4 x 10
5 x 0	5 x 1	5 x 2	5 x 3	5 x 4	5 x 5	5 x 6	5 x 7	5 x 8	5 x 9	5 x 10
6 x 0	6 x 1	6 x 2	6 x 3	6 x 4	6 x 5	6 x 6	6 x 7	6 x 8	6 x 9	6 x 10
7 x 0	7 x 1	7 x 2	7 x 3	7 x 4	7 x 5	7 x 6	7 x 7	7 x 8	7 x 9	7 x 10
8 x 0	8 x 1	8 x 2	8 x 3	8 x 4	8 x 5	8 x 6	8 x 7	8 x 8	8 x 9	8 x 10
9 x 0	9 x 1	9 x 2	9 x 3	9 x 4	9 x 5	9 x 6	9 x 7	9 x 8	9 x 9	9 x 10
10 x 0	10 x 1	10 x 2	10 x 3	10 x 4	10 x 5	10 x 6	10 x 7	10 x 8	10 x 9	10 x 10

5¢ = 1 Nickel - A good place to apply the five facts is with money. We know that 5¢ is the same as one nickel, so we can ask how many pennies in four nickels to apply 4 x 5. The answer is 20¢.

1 Nickel   5¢   1¢   1¢   1¢   1¢   1¢

Example 2

How many pennies, or cents, in four nickels?

5¢   5¢   5¢   5¢   4 • 5¢ = 20¢

LESSON 7

Area of a Rectangle and a Square

Up till now, we have always had squares inside the rectangles to show the area. The rectangle is the way we illustrate multiplication. The over dimension and the up dimension are the factors, and the area is the product. Multiplication of the factors gives the product. We will be applying our multiplication skills to find the area of rectangles when given the dimensions in inches (") and feet ('). The answers to an area question are always in terms of square feet, square inches, or square units if a unit of measure has not been given. To help the students remember this, I use the word "squarea." This combines the two words "square" and "area."

Study the examples to make sure you understand how to find the area of a rectangle or the special rectangle called a square. When in doubt, build the rectangle with the blocks to see how many "square" units there are.

Example 1

Find the area of the rectangle.

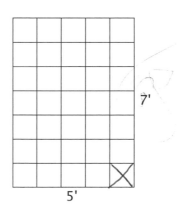

Area = (5 ft)(7 ft) = 35 square feet
or 35 sq ft

Example 2

Find the area of the rectangle.

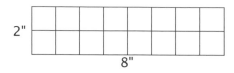

Area = (8 in)(2 in) = 16 square inches
or 16 sq in

Example 3

Find the area of the square.

10

10

Area = (10)(10) = 100 square units
or 100 sq units

When a unit of measure is not specified, label your answer with the generic term *units*.

Example 4

Find the area of the square.

5"

5"

Area = (5 in)(5 in) = 25 square inches
or 25 sq in

Because the figure in example 4 is a square, if we know the length of one side, we know the length of all the sides. A square is a rectangle with all the sides the same length.

LESSON 8

Solve for an Unknown

When learning the multiplication facts, we also like to solve for an unknown. There are three reasons why we introduce this topic now. Solving for an unknown reviews multiplication, introduces algebra taught concretely, and lays a solid foundation for division. This is not a light topic. It is very important to our next unit of study, which is division. While solving for an unknown is laying a foundation for the formal study of algebra, that is still several years off; but how you well you teach this subject now will have impact on learning single digit division.

In multiplication, you are given the two factors, and you have to find the product. In division you are given the product and one factor, and you have to find the missing factor. This is exactly what we are doing in solving for an unknown.

How you verbalize the equation can be the key to understanding what it means. $2G = 12$ can be read as, "Two counted how many times is twelve?" or, "What number counted two times is twelve?" This is because of the commutative property of multiplication, which states that you can change the order of the factors. You might also verbalize this as, "Two times what equals twelve?" or, "What times two equals twelve?" Either way is acceptable. Choose the way that is easiest for your student to understand. The first example illustrates this with the manipulative blocks. Study these examples until this important concept is understood.

Example 1

Solve for the unknown, or find out the value of "G" in 2G = 12.

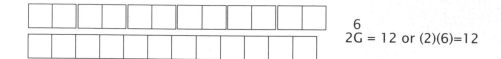

6
2G = 12 or (2)(6)=12

I solved it as, "How many twos can I count out of twelve?" or, "What times two equals twelve?" Using the blocks, you can see that there are six twos in twelve. So 2 x 6 = 12, and the missing factor is 6. When you find the answer, simply write a 6 just above the G. Later on, in *Pre-Algebra*, we'll begin writing another line below the problem, with the solution written as G = 6.

Example 2

Solve for the unknown, or find out the value of "X" in 5X = 15.

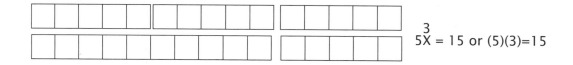

3
5X = 15 or (5)(3)=15

I solved it as, "How many fives can I count out of fifteen?" or, "What times five equals fifteen?" Using the blocks, you can see that there are three fives in fifteen. So 5 x 3 = 15, and the missing factor is 3. When you find the answer, simply write a 3 just above the X.

Skip Count by 9, Equivalent Fractions

On this page, we are skip counting by nine. Use the same techniques to introduce and teach this important skill that have worked before. A practical example is nine players on a baseball team.

Example 1

Skip count and write the numbers on the lines. Say each number out loud as you count and write. After filling out the rectangle, write the numbers in the spaces provided beneath the figure. The solution is below.

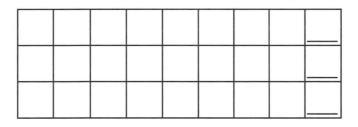

9 , ___ , ___

Solution

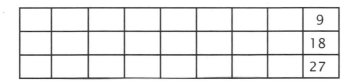

9 , 18 , 27

Multiples in Equivalent Fractions - In lesson 3, one of the reasons cited for learning the skip-count facts was the advantage it gives you when learning fractions. In the worksheets, when reinforcing skip-counting facts, we will give the problem in fraction form and ask the student to fill in the multiples of the numbers.

In example 2, we see the multiples of 2/5. Lesson 13 will have more on fractions.

Example 2

Find the missing multiples of 2 and 5 in the equivalent fractions.

$$\frac{2}{5} = \frac{4}{\quad} = \frac{6}{15} = \frac{\quad}{\quad} = \frac{\quad}{25} = \frac{12}{\quad} = \frac{\quad}{35} = \frac{\quad}{\quad} = \frac{\quad}{\quad} = \frac{20}{50}$$

Solution

$$\frac{2}{5} = \frac{4}{10} = \frac{6}{15} = \frac{8}{20} = \frac{10}{25} = \frac{12}{30} = \frac{14}{35} = \frac{16}{40} = \frac{18}{45} = \frac{20}{50}$$

Multiply by 9

Multiplication by nine has a unique pattern. I show this on the DVD and will try to explain it here as well.

Begin with nine green units in the units place. That is 1 x 9 or 9. Now to show 2 x 9, I need to add another nine, but the units place is full, so I add a ten, which is one too many, and take one away from the units place. By adding one to the tens place and subtracting one from the units place, I still have nine blocks on the board. The answer for 2 x 9 is 18. By adding another ten to the tens place and taking another unit from the units place, I now have three nines, or 3 x 9, which is 27. Continuing on by adding ten and subtracting one (the same as adding a nine), I proceed through all of the multiples of nine: 9–18–27–36–45–54–63–72–81–90. Because we took away a unit whenever we added a ten, we always added nine blocks. This reveals the interesting pattern. Look at the individual digits. Now add them! For 18, 1 + 8 = 9, for 27, 2 + 7 = 9, and for 36, 3 + 6 = 9.

Get the idea? We can tell whether a number is a multiple of nine simply by adding the digits, and seeing if they will add up to nine or a multiple of nine. Example: for 144, 1 + 4 + 4 = 9. Sure enough, 9 x 16 = 144. So 144 is a multiple of nine. Now let's relate this to the 10 facts. We know 10 x 6 is 60, so 9 x 6 will be a little less, right? Let's say 50 something or 5___. What plus 5 makes 9? The answer is 4! So, 9 x 6 = 54. Do several of these to discover a pattern. We took one less than six for the number in the tens place (five), and then found what added to five makes nine (four). The final answer is 54.

9 x 8   The tens place is one less than 8, which is 7, and 7 + 2 = 9,
          so the answer is 72!

9 x 4   The tens place is one less than 4, which is 3, and 3 + 6 = 9,
          so the answer is 36!

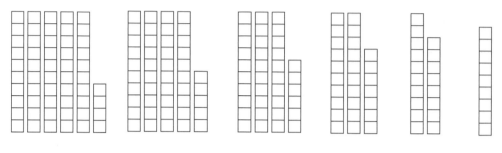

9·6 = 54   9·5 = 45   9·4 = 36   9·3 = 27   9·2 = 18   9·1 = 9

Another way to show this is on a number chart. Circling all of the nine facts, or multiples of nine, reveals the pattern that corresponds to the blocks above.

0	1	2	3	4	5	6	7	8	⑨
10	11	12	13	14	15	16	17	⑱	19
20	21	22	23	24	25	26	㉗	28	29
30	31	32	33	34	35	㊱	37	38	39
40	41	42	43	44	㊺	46	47	48	49
50	51	52	53	㊴	55	56	57	58	59
60	61	62	㊿	64	65	66	67	68	69
70	71	㊲	73	74	75	76	77	78	79
80	㊶	82	83	84	85	86	87	88	89
⑨⓪									

Of course, each fact can be built in the shape of a rectangle. Whenever illustrating with the blocks, also write it and say it as you build.

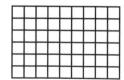

Nine counted six times is the same as 54,
or 9 times 6 equals 54,
or 9 over and 6 up is 54.

On the worksheets, there have been rectangles where the student wrote in the fact at the end of the line, in the space with an underline. These can be put to another use by adding the multiplication problem to the multiple of nine. Here are a few examples.

| | | | | | | | | 9 | Nine counted one time equals nine or 9 x 1 = 9.

| | | | | | | | | 18 | Nine counted two times equals eighteen or 9 x 2 = 18.

| | | | | | | | | 27 | Nine counted three times equals twenty-seven or 9 x 3 = 27.

| | | | | | | | | 36 | Nine counted four times equals thirty-six or 9 x 4 = 36.

Skip counting by nine is the first step. After this is accomplished, say the factors slowly, and then ask the student to say the product. For example, you say "nine counted one time" or "nine times one" and the student says "nine." Continue by saying "nine times two" and the student says "18." Proceed through all the facts sequentially, just like the student learned to skip count by nine.

Here are the nine facts with the corresponding numbering.

0	9	18	27	36	45	54	63	72	81	90
(9)(0)	(9)(1)	(9)(2)	(9)(3)	(9)(4)	(9)(5)	(9)(6)	(9)(7)	(9)(8)	(9)(9)	(9)(10)

↑   ↑   ↑

9 counted 0 times   9 counted 5 times   9 counted 8 times

0 x 0	0 x 1	0 x 2	0 x 3	0 x 4	0 x 5	0 x 6	0 x 7	0 x 8	0 x 9	0 x 10
1 x 0	1 x 1	1 x 2	1 x 3	1 x 4	1 x 5	1 x 6	1 x 7	1 x 8	1 x 9	1 x 10
2 x 0	2 x 1	2 x 2	2 x 3	2 x 4	2 x 5	2 x 6	2 x 7	2 x 8	2 x 9	2 x 10
3 x 0	3 x 1	3 x 2	3 x 3	3 x 4	3 x 5	3 x 6	3 x 7	3 x 8	3 x 9	3 x 10
4 x 0	4 x 1	4 x 2	4 x 3	4 x 4	4 x 5	4 x 6	4 x 7	4 x 8	4 x 9	4 x 10
5 x 0	5 x 1	5 x 2	5 x 3	5 x 4	5 x 5	5 x 6	5 x 7	5 x 8	5 x 9	5 x 10
6 x 0	6 x 1	6 x 2	6 x 3	6 x 4	6 x 5	6 x 6	6 x 7	6 x 8	6 x 9	6 x 10
7 x 0	7 x 1	7 x 2	7 x 3	7 x 4	7 x 5	7 x 6	7 x 7	7 x 8	7 x 9	7 x 10
8 x 0	8 x 1	8 x 2	8 x 3	8 x 4	8 x 5	8 x 6	8 x 7	8 x 8	8 x 9	8 x 10
9 x 0	9 x 1	9 x 2	9 x 3	9 x 4	9 x 5	9 x 6	9 x 7	9 x 8	9 x 9	9 x 10
10 x 0	10 x 1	10 x 2	10 x 3	10 x 4	10 x 5	10 x 6	10 x 7	10 x 8	10 x 9	10 x 10

Skip Count by 3

Before we learn how to multiply by three, we must master skip counting by three. If the student learns skip counting by singing songs, make sure he can also say them in a monotone voice without the aid of music. Some ideas for making this skill practical are the sides of a triangle or the wheels on a tricycle.

Example 1

Skip count and write the numbers on the lines. Say each number out loud as you count and write. After filling out the rectangle, write the numbers in the spaces provided beneath the figure.

		3
		6
		9
		12
		15
		18

Notice that this rectangle illustrates a specific multiplication problem. It is 3 x 6 = 18.

3,   6,   9,   12,   15,   18

Let's practice this new skill by counting the sides of a triangle and the dots on one side of a domino.

Example 2

How many sides on four triangles?

"3-6-9-12." There are 12 sides on four triangles.

Example 3

How many dots on six sides?

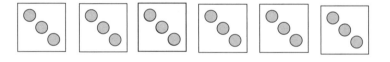

"3-6-9-12-15-18." There are 18 dots on six sides.

Multiply by 3, 3 Feet = 1 Yard
1 Tablespoon = 3 Teaspoons

Now we have learned the 0, 1, 2, 5, 9, and 10 facts. The small chart on the next page shows that we have only the following multiplication by three facts to learn: 3 x 3, 3 x 4, 3 x 6, 3 x 7, and 3 x 8. The other three facts have been learned while multiplying by the other numbers. Be encouraged; we are over halfway done.

Use all of the techniques we have learned in the past for learning the facts. When all else fails, simply memorize them. All of the aids and patterns are helps to learning, but it is very important that the student(s) have their facts down cold before moving any further. Notice the products of the threes. The digits in the products add up to three or a multiple of three. For example, for 21, 2 + 1 = 3, for 24, 2 + 4 = 6, and for 27, 2 + 7 = 9. This is similar to the pattern for the nines.

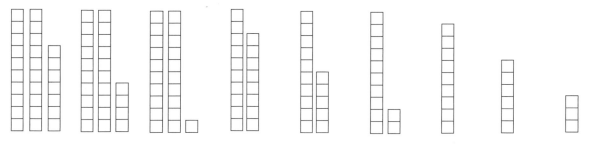

$3 \cdot 9 = 27$ $3 \cdot 8 = 24$ $3 \cdot 7 = 21$ $3 \cdot 6 = 18$ $3 \cdot 5 = 15$ $3 \cdot 4 = 12$ $3 \cdot 3 = 9$ $3 \cdot 2 = 6$ $3 \cdot 1 = 3$

Another way to show this is on a number chart. Circling all of the three facts, or multiples of three, reveals an interesting pattern that corresponds to the blocks on the previous page.

0   1   2   ③   4   5   ⑥   7   8   ⑨
10 11 ⑫ 13 14 ⑮ 16 17 ⑱ 19
20 ㉑ 22 23 ㉔ 25 26 ㉗ 28 29

Since the student has learned the skip-counting facts, he or she knows all of the multiples of three, or products of three. Say the factors slowly, and then ask the student to say the product, which is skip counting. You say, "three counted one time" or "three times one" and the student says "three." You continue by saying "three times two" and the student says "six." Proceed through all the facts sequentially, just as the student learned the skip counting facts.

Here are the 3 facts with the corresponding numbering.

$\underline{0}$	$\underline{3}$	$\underline{6}$	$\underline{9}$	$\underline{12}$	$\underline{15}$	$\underline{18}$	$\underline{21}$	$\underline{24}$	$\underline{27}$	$\underline{30}$
3×0	3×1	3×2	3×3	3×4	3×5	3×6	3×7	3×8	3×9	3×10

↑   ↑   ↑
3 counted 2 times   3 counted 6 times   3 counted 10 times

Of course each fact can be built in the shape of a rectangle. Whenever illustrating with the blocks, also write it and say it as you build.

3 counted four times is the same as 12, or
3 times 4 equals 12, or
3 over and 4 up is 12.

On the worksheets, there have been rectangles where the student wrote in the fact at the end of the line in the space with an underline. These can be put to another use by adding the multiplication problem to the multiple of three. Here are a few examples.

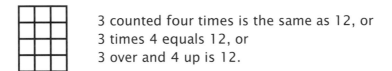

Three counted one time equals three or 3 x 1 = 3.
Three counted two times equals six or 3 x 2 = 6.
Three counted three times equals nine or 3 x 3 = 9.

3 Feet = 1 Yard - Show the student(s) a one-foot ruler and a three-foot yard-stick. Hold up the ruler and teach them this is one foot long. It measures length, which is how long something is. Measure some people, and see how many feet tall they are. Measure how far to the ceiling and how wide their desks or tables are, etc. Remember the symbol for feet is **'**. Four feet is the same as 4'. Six feet is 6'.

Then introduce the yardstick, and using the ruler, show that it is the same as three rulers or three feet. Then measure objects with the yardstick, like the length of the room. If the students are interested in football, show them five yards (the distance between stripes on a football field), and then show 10 yards (what is needed for a first down). If they like track, mention the 100-yard dash and the 40-yard dash.

After they are familiar with the yardstick, tell them that the man who wrote this book is around two yards tall. Then ask them how many feet that represents. If there are three feet in one yard, then there are three plus three or six feet in two yards. This an appropriate place to apply multiplying by threes as we change yards to feet.

1 yard

| 1' | 1' | 1' |

Example 1

How many feet are there in two yards?

2 yards = 2 x 3' (or 1 yard) = 6'

Liquid Measure - Teaspoons and Tablespoons - Begin with a one-tea-spoon measuring spoon and a one-tablespoon measuring spoon . Fill up the one-teaspoon measuring spoon and empty it into the tablespoon. Do this three times. The student sees that one tablespoon equals three teaspoons.

Example 2

How many teaspoons are in four tablespoons?

4 Tbsp = 4 x 3 (tsp in a Tbsp) = 12 tsp

1 tablespoon   =   3
                    teaspoons

Abbreviations:

Tablespoon = Tbsp
teaspoon = tsp

0 x 0	0 x 1	0 x 2	0 x 3	0 x 4	0 x 5	0 x 6	0 x 7	0 x 8	0 x 9	0 x 10
1 x 0	1 x 1	1 x 2	1 x 3	1 x 4	1 x 5	1 x 6	1 x 7	1 x 8	1 x 9	1 x 10
2 x 0	2 x 1	2 x 2	2 x 3	2 x 4	2 x 5	2 x 6	2 x 7	2 x 8	2 x 9	2 x 10
3 x 0	3 x 1	3 x 2	3 x 3	3 x 4	3 x 5	3 x 6	3 x 7	3 x 8	3 x 9	3 x 10
4 x 0	4 x 1	4 x 2	4 x 3	4 x 4	4 x 5	4 x 6	4 x 7	4 x 8	4 x 9	4 x 10
5 x 0	5 x 1	5 x 2	5 x 3	5 x 4	5 x 5	5 x 6	5 x 7	5 x 8	5 x 9	5 x 10
6 x 0	6 x 1	6 x 2	6 x 3	6 x 4	6 x 5	6 x 6	6 x 7	6 x 8	6 x 9	6 x 10
7 x 0	7 x 1	7 x 2	7 x 3	7 x 4	7 x 5	7 x 6	7 x 7	7 x 8	7 x 9	7 x 10
8 x 0	8 x 1	8 x 2	8 x 3	8 x 4	8 x 5	8 x 6	8 x 7	8 x 8	8 x 9	8 x 10
9 x 0	9 x 1	9 x 2	9 x 3	9 x 4	9 x 5	9 x 6	9 x 7	9 x 8	9 x 9	9 x 10
10 x 0	10 x 1	10 x 2	10 x 3	10 x 4	10 x 5	10 x 6	10 x 7	10 x 8	10 x 9	10 x 10

Skip Count by 6, Equivalent Fractions

Skip counting by six will give us a head start in learning to multiply by six. Even after we learn our facts, skip counting can still aid us. There are times when I have seen a student forget that 6 x 6 = 36. Instead of panicking, a student who knows how to skip count and understands that skip counting is fast adding of the same number can figure it out. He or she might say, "Well, I know that 5 x 6 = 30, so 6 x 6 must be 6 more than that or 36." Reinforce this as you learn these facts in order. If the student learns them by singing songs, make sure he can also skip count in a monotone voice, without the aid of music.

Some ideas for making this skill practical are the sides of a hexagon, the legs of an insect, a half dozen, the walls in a honeycomb, and the dots on a die (singular for dice).

Example 1

Skip count and write the numbers on the lines. Say each number out loud as you count and write. After filling out the rectangle, write the numbers in the spaces provided beneath the figure.

					6
					12
					18
					24
					30
					36

Notice that this rectangle illustrates a specific problem. It is 6 x 6 = 36.

6,   12,   18,   24,   30,   36

Fractions - For the last few lessons, we have been finding the multiples, or skip-counting facts, in fraction form. In this lesson, we will show what a fraction and an equivalent fraction are using rectangles. A *fraction* has a numerator and a denominator. The *denominator* is the number that represents how many total parts there are and is written at the bottom of a fraction. The number of the total parts that are shaded is called the "numberator" or *numerator*. In the next few examples, first count the total parts and write that number at the bottom, and then count how many are shaded and write that number on the top.

Example 2

Find the denominator and numerator of the rectangle.

$$\frac{\text{numerator}}{\text{denominator}} = \frac{3}{5} = \frac{\text{how many}}{\text{total parts}}$$

Example 3

Find the denominator and numerator of the rectangle.

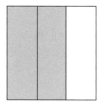

$$\frac{\text{numerator}}{\text{denominator}} = \frac{2}{3} = \frac{\text{how many}}{\text{total parts}}$$

NOTE: With the teaching on the DVD using the overlays and the pictures on the worksheets, the student shouldn't need to have his own set to solve the problems. Wait until the book on fractions before purchasing the fraction overlays. At this level, cooking and other real-life applications of fractions are recommended.

Equivalent Fractions - An **equivalent fraction** is a fraction with the same amount but more pieces. Think of example 3 as being a cake that was cut into three pieces. Then one of the pieces was eaten. There are two pieces left, but several people come to eat the cake, so they have to divide it into more pieces. It is still the same amount of cake (2/3), but there are more pieces, as shown in example 4. Notice that the "numberator" and denominator of equivalent fractions are multiples that we find as a result of skip counting.

Example 4

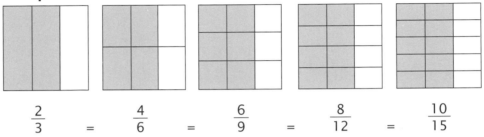

$$\frac{2}{3} \quad = \quad \frac{4}{6} \quad = \quad \frac{6}{9} \quad = \quad \frac{8}{12} \quad = \quad \frac{10}{15}$$

Example 5

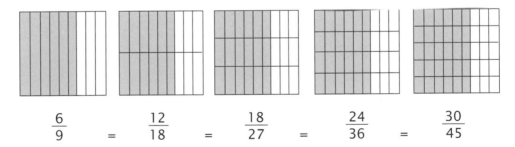

$$\frac{6}{9} \quad = \quad \frac{12}{18} \quad = \quad \frac{18}{27} \quad = \quad \frac{24}{36} \quad = \quad \frac{30}{45}$$

Example 6

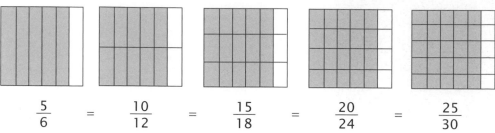

$$\frac{5}{6} = \frac{10}{12} = \frac{15}{18} = \frac{20}{24} = \frac{25}{30}$$

Multiply by 6

There are a few patterns that emerge as we study the six facts. But remember, we have only four of these facts left to learn: 6 x 4, 6 x 6, 6 x 7, and 6 x 8. Since six is a multiple of three and a multiple of two, in all the answers to the sixes the digits will add up to a multiple of three and be even. Recently, someone noticed that when multiplying six by an even number, the answer always ends in the same number as you multiplied by.

Notice 6 x 4 = 24. The answer ends in four. Then observe that the number in the tens place is half of the number you multiplied by, or two. Try it with the other even multiples. The pattern is based on what we learned about the fives. Multiplying 5 x 4 gives us 20, and 1 x 4 gives us 4. When we multiplied by five we took half of the number, and then added a zero.

Using this pattern to multiply by four, you take half of the number then add the number. For 6 x 4, take half of the number four, which is two, and then add the number four. So the answer is 24. To show this, consider that 6 is 5 + 1, and look at the picture below. If this is confusing, don't worry about it, simply employ whatever strategy got you this far and learn the six facts.

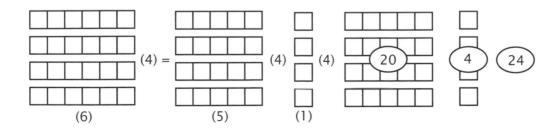

On the worksheets, there have been rectangles where the student wrote in the fact at the end of the line, in the space with an underline. These can be put to another use by adding the multiplication problem to the multiple of six.

	6	Six counted one time equals six or 6 x 1 = 6.
	12	Six counted two times equals twelve or 6 x 2 = 12.
	18	Six counted three times equals eighteen or 6 x 3 = 18.
	24	Six counted four times equals twenty-four or 6 x 4 = 24.
	30	Six counted five times equals thirty or 6 x 5 = 30.
	36	Six counted six times equals thirty-six or 6 x 6 = 36.
	42	Six counted seven times equals forty-two or 6 x 7 = 42.
	48	Six counted eight times equals forty-eight or 6 x 8 = 48.

0 x 0	0 x 1	0 x 2	0 x 3	0 x 4	0 x 5	0 x 6	0 x 7	0 x 8	0 x 9	0 x 10
1 x 0	1 x 1	1 x 2	1 x 3	1 x 4	1 x 5	1 x 6	1 x 7	1 x 8	1 x 9	1 x 10
2 x 0	2 x 1	2 x 2	2 x 3	2 x 4	2 x 5	2 x 6	2 x 7	2 x 8	2 x 9	2 x 10
3 x 0	3 x 1	3 x 2	3 x 3	3 x 4	3 x 5	3 x 6	3 x 7	3 x 8	3 x 9	3 x 10
4 x 0	4 x 1	4 x 2	4 x 3	4 x 4	4 x 5	4 x 6	4 x 7	4 x 8	4 x 9	4 x 10
5 x 0	5 x 1	5 x 2	5 x 3	5 x 4	5 x 5	5 x 6	5 x 7	5 x 8	5 x 9	5 x 10
6 x 0	6 x 1	6 x 2	6 x 3	6 x 4	6 x 5	6 x 6	6 x 7	6 x 8	6 x 9	6 x 10	←
7 x 0	7 x 1	7 x 2	7 x 3	7 x 4	7 x 5	7 x 6	7 x 7	7 x 8	7 x 9	7 x 10
8 x 0	8 x 1	8 x 2	8 x 3	8 x 4	8 x 5	8 x 6	8 x 7	8 x 8	8 x 9	8 x 10
9 x 0	9 x 1	9 x 2	9 x 3	9 x 4	9 x 5	9 x 6	9 x 7	9 x 8	9 x 9	9 x 10
10 x 0	10 x 1	10 x 2	10 x 3	10 x 4	10 x 5	10 x 6	10 x 7	10 x 8	10 x 9	10 x 10

↑

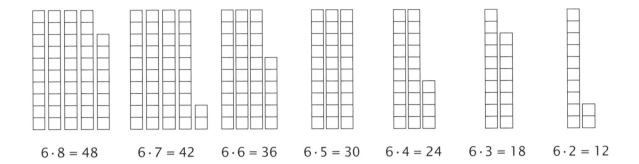

$6 \cdot 8 = 48 \qquad 6 \cdot 7 = 42 \qquad 6 \cdot 6 = 36 \qquad 6 \cdot 5 = 30 \qquad 6 \cdot 4 = 24 \qquad 6 \cdot 3 = 18 \qquad 6 \cdot 2 = 12$

Another way to show this is on a number chart. Circling all of the six facts, or multiples of six, reveals an interesting pattern that corresponds to the blocks above.

```
 0  1  2  3  4  5 (6) 7  8  9
10 11(12)13 14 15 16 17(18)19
20 21 22 23(24)25 26 27 28 29
(30)31 32 33 34 35(36)37 38 39
40 41(42)43 44 45 46 47(48)49
50 51 52 53(54)55 56 57 58 59
(60)61 62 63 64 65 66 67 68 69
```

Since the student has learned the skip counting facts and is able to say them without music, he or she knows all of the multiples of six or products of six. This is a good time for you as the teacher to put the factors with the product. Say the factors slowly, and then ask the student to say the product, which is the skip counting. For example, you say "six counted one time" or "six times one" and the student says "six." You continue by saying "six times two" and the student says "twelve." Proceed through all the facts sequentially, just as he or she learned the skip-counting facts.

Here are the six facts with the corresponding numbering.

| $\dfrac{0}{6\times0}$ | $\dfrac{6}{6\times1}$ | $\dfrac{12}{6\times2}$ | $\dfrac{18}{6\times3}$ | $\dfrac{24}{6\times4}$ | $\dfrac{30}{6\times5}$ | $\dfrac{36}{6\times6}$ | $\dfrac{42}{6\times7}$ | $\dfrac{48}{6\times8}$ | $\dfrac{54}{6\times9}$ | $\dfrac{60}{6\times10}$ |

↑   ↑   ↑

6 counted 3 times   6 counted 6 times   6 counted 9 times

Of course each fact can be built in the shape of a rectangle. Whenever illustrating with the blocks, also write it and say it as you build.

6 counted six times is the same as 36, or
6 times 6 equals 36, or
6 over and 6 up is 36.

Skip Count by 4, 4 Quarts = 1 Gallon

After we know how to skip count or fast add by four, we can apply this to counting how many quarts in gallons. If you have three gallons, then you would have 4–8–12 quarts in three gallons. While learning these facts, apply them to the liquid measure of four quarts in one gallon.

Example 1

Skip count and write the numbers on the lines. Say each number out loud as you count and write. After filling out the rectangle, write the numbers in the spaces provided beneath the figure.

			4
			8
			12
			16

Notice that this rectangle illustrates a specific multiplication problem. It is 4 x 4 = 16.

_4_ , _8_ , _12_ , _16_

4 Quarts = 1 Gallon - If you can illustrate this with liquid, begin with four one-quart containers and one one-gallon container. Fill up the one-gallon container, and empty it into the four one-quart containers. The student can see that one gallon equals four quarts. Or do the converse by emptying the four quarts into the one gallon.

Example 2

How many quarts in two gallons

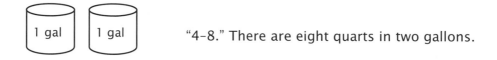

"4–8." There are eight quarts in two gallons.

Example 3

How many quarts in four gallons?

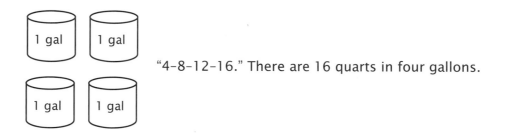

"4-8-12-16." There are 16 quarts in four gallons.

Multiply by 4, 4 Quarters = 1 Dollar

No tricks or patterns here. Just learn the ones you don't know: 4 x 4, 4 x 7, and 4 x 8, and then review the ones you do.

4	Four counted one time equals four or 4 x 1 = 4.
8	Four counted two times equals eight or 4 x 2 = 8.
12	Four counted three times equals twelve or 4 x 3 = 12.
16	Four counted four times equals sixteen or 4 x 4 = 16.
20	Four counted five times equals twenty or 4 x 5 = 20.
24	Four counted six times equals twenty-four or 4 x 6 = 24.
28	Four counted seven times equals twenty-eight or 4 x 7 = 28.
32	Four counted eight times equals thirty-two or 4 x 8 = 32.

$\frac{0}{4 \times 0}$	$\frac{4}{4 \times 1}$	$\frac{8}{4 \times 2}$	$\frac{12}{4 \times 3}$	$\frac{16}{4 \times 4}$	$\frac{20}{4 \times 5}$	$\frac{24}{4 \times 6}$	$\frac{28}{4 \times 7}$	$\frac{32}{4 \times 8}$	$\frac{36}{4 \times 9}$	$\frac{40}{4 \times 10}$
				↑			↑			
				4 counted 4 times			4 counted 7 times			

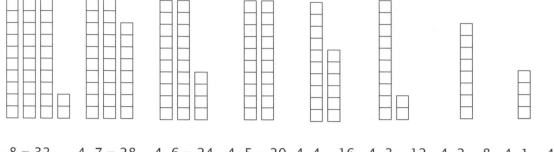

$4 \cdot 8 = 32$ $4 \cdot 7 = 28$ $4 \cdot 6 = 24$ $4 \cdot 5 = 20$ $4 \cdot 4 = 16$ $4 \cdot 3 = 12$ $4 \cdot 2 = 8$ $4 \cdot 1 = 4$

Another way to show this is on a number chart. Circling all of the four facts, or multiples of four, reveals an interesting pattern that corresponds to the blocks above.

0 1 2 3 ④ 5 6 7 ⑧ 9
10 11 ⑫ 13 14 15 ⑯ 17 18 19
⑳ 21 22 23 ㉔ 25 26 27 ㉘ 29
30 31 ㉜ 33 34 35 ㊱ 37 38 39
㊵

1 Quarter = 25¢ and 4 Quarters = 1 Dollar - Hold up a quarter and teach that it is one quarter or 25 cents. We write it as 25¢. With the blocks, show one quarter by holding up two ten bars and one five bar.

25¢ = "twenty-five cents" = [quarter] = [blocks]

Using the blocks, discover how many quarters are in one dollar (100 block). Then hold up a dollar bill in one hand and four quarters in the other to show that these are the same amount.

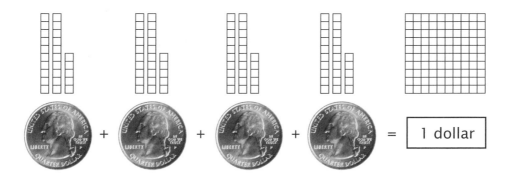

So four quarters = one dollar.

Example

How many quarters in three dollars?

= 3 x 4 =12 quarters

Of course, each fact can be built in the shape of a rectangle. Whenever illustrating with the blocks, also write it and say it as you build.

 4 counted four times is the same as 16, or
4 times 4 equals 16, or
4 over and 4 up is 16.

0 x 0	0 x 1	0 x 2	0 x 3	0 x 4	0 x 5	0 x 6	0 x 7	0 x 8	0 x 9	0 x 10
1 x 0	1 x 1	1 x 2	1 x 3	1 x 4	1 x 5	1 x 6	1 x 7	1 x 8	1 x 9	1 x 10
2 x 0	2 x 1	2 x 2	2 x 3	2 x 4	2 x 5	2 x 6	2 x 7	2 x 8	2 x 9	2 x 10
3 x 0	3 x 1	3 x 2	3 x 3	3 x 4	3 x 5	3 x 6	3 x 7	3 x 8	3 x 9	3 x 10
4 x 0	4 x 1	4 x 2	4 x 3	4 x 4	4 x 5	4 x 6	4 x 7	4 x 8	4 x 9	4 x 10
5 x 0	5 x 1	5 x 2	5 x 3	5 x 4	5 x 5	5 x 6	5 x 7	5 x 8	5 x 9	5 x 10
6 x 0	6 x 1	6 x 2	6 x 3	6 x 4	6 x 5	6 x 6	6 x 7	6 x 8	6 x 9	6 x 10
7 x 0	7 x 1	7 x 2	7 x 3	7 x 4	7 x 5	7 x 6	7 x 7	7 x 8	7 x 9	7 x 10
8 x 0	8 x 1	8 x 2	8 x 3	8 x 4	8 x 5	8 x 6	8 x 7	8 x 8	8 x 9	8 x 10
9 x 0	9 x 1	9 x 2	9 x 3	9 x 4	9 x 5	9 x 6	9 x 7	9 x 8	9 x 9	9 x 10
10 x 0	10 x 1	10 x 2	10 x 3	10 x 4	10 x 5	10 x 6	10 x 7	10 x 8	10 x 9	10 x 10

Skip Count by 7, Multiples of 10

Before multiplying by seven, skip counting by sevens will give us our foundation. Use the same techniques to introduce these facts that have worked before. Don't forget, if the student learns them by singing songs, make sure he or she can also say them in a monotone voice, without the aid of music.

Example 1

Skip count and write the number in the boxes with the lines. Whisper as you count the squares, and then say the last number out loud as you write it.

						7
						14
						21
						28
						35
						42
						49
						56

Notice that this rectangle illustrates a specific problem. It is 7 x 8 = 56.

__7__, __14__, __21__, __28__, __35__, __42__, __49__, __56__

Multiplication by Multiples of 10 - When multiplying by multiples of 10 such as 20, 30, or 40, I like to multiply the factor times the digit in the tens place, and then add the zero from the units place. To do this, I have a mitten that represents the teacher's hand to cover the zero when initially multiplying. Then I remove the mitten and add the zero.

Example 2

Multiply 20 times 4.

Example 3

Multiply 30 times 2.

Multiply by 7 and by Multiples of 100

There is one trick for 7 x 8. Write the numbers in sequence as 5–6–7–8. Do you see the multiplication problem? Supply an equal sign and a times symbol and you have 56 = 7 x 8. The only facts in this lesson that you have not been exposed to before are 7 x 7 and 7 x 8. Review all the seven facts that you have previously learned and focus on the two new ones before moving to the next lesson.

7	Seven counted one time equals seven or 7 x 1 = 7.
14	Seven counted two times equals fourteen or 7 x 2 = 14.
21	Seven counted three times equals twenty-one or 7 x 3 = 21.
28	Seven counted four times equals twenty-eight or 7 x 4 = 28.
35	Seven counted five times equals thirty-five or 7 x 5 = 35.
42	Seven counted six times equals forty-two or 7 x 6 = 42.

0	7	14	21	28	35	42	49	56	63	70
7×0	7×1	7×2	7×3	7×4	7×5	7×6	7×7	7×8	7×9	7×10

        ↑   ↑

7 counted 3 times   7 counted 8 times

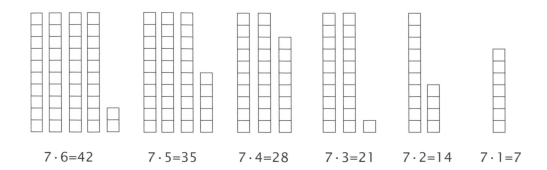

7·6=42   7·5=35   7·4=28   7·3=21   7·2=14   7·1=7

Another way to show this is on a number chart. Here we circle all of the seven facts or multiples of seven.

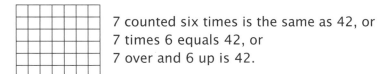

```
 0    1    2    3   (4)   5    6   (7)   8    9
10   11   12   13  (14)  15   16   17   18   19
20  (21)  22   23   24   25   26   27  (28)  29
30   31   32   33   34  (35)  36   37   38   39
40   41  (42)  43   44   45   46   47   48  (49)
50   51   52   53   54   55  (56)  57   58   59
60   61   62  (63)  64   65   66   67   68   69
(70)
```

Of course each fact can be built in the shape of a rectangle. Whenever illustrating with the blocks, also write it and say it as you build.

7 counted six times is the same as 42, or
7 times 6 equals 42, or
7 over and 6 up is 42.

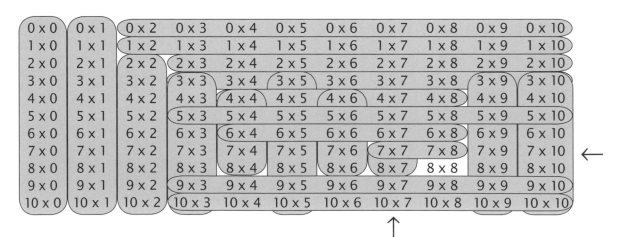

Days in a Week - The most common application of multiplying by seven is the days in a week.

Example 1

How many days are in six weeks? 7 x 6 = 42 days

Example 2

How many days are in seven weeks? 7 x 7 = 49 days

Multiplication by Multiples of 100 - When multiplying by multiples of 100, I like to multiply the factor times the digit in the hundreds place and then add the two zeros from the tens and units places. To do this, I have a mitten that represents the teacher's hand to cover the zeros when initially multiplying. Then I remove the mitten and add the zeros.

Example 3

Multiply 200 times 4.

$$
\begin{array}{r} 200 \\ \times\ 4 \\ \hline \end{array}
\longrightarrow
\begin{array}{r} 2 \\ \times \\ \hline 8 \end{array}
\longrightarrow
\begin{array}{r} 200 \\ \times\ 4 \\ \hline 800 \end{array}
$$

Example 4

Multiply 300 times 2.

$$
\begin{array}{r} 300 \\ \times\ 2 \\ \hline \end{array}
\longrightarrow
\begin{array}{r} 3 \\ \times \\ \hline 6 \end{array}
\longrightarrow
\begin{array}{r} 300 \\ \times\ 2 \\ \hline 600 \end{array}
$$

Example 5

Multiply 100 times 8.

$$\begin{array}{r} 100 \\ \times\ 8 \\ \hline \end{array} \longrightarrow \begin{array}{r} 1 \\ \times\ 8 \\ \hline 8 \end{array} \longrightarrow \begin{array}{r} 100 \\ \times\ 8 \\ \hline 800 \end{array}$$

Skip Count by 8, 8 Pints = 1 Gallon

Before we can multiply by eight, skip counting by eight will give us a head start. Use the same techniques to introduce these facts that have worked before. Don't forget, if the student learns them by singing songs, make sure he or she can also say them in a monotone voice without the aid of music.

Practical examples for the eights are sides on a stop sign, pints in a gallon, legs on a spider, and arms on an octopus.

"Octo" means eight. An *octagon* has eight sides, and an octopus has eight arms.

Example 1

Skip count and write the number in the boxes with the lines. Whisper as you count the squares, and then say the last number out loud as you write it.

After filling out the rectangle write the numbers in the spaces provided beneath the figure.

							8
							16
							24
							32
							40
							48
							56
							64

Notice that this rectangle illustrates a specific problem. It is 8 x 8 = 64.

_8_ , _16_ , _24_ , _32_ , _40_ , _48_ , _56_ , _64_

LESSON 20

Multiply by 8

There is only one fact you haven't seen, and that is 8 x 8. Review all the eight facts that you have previously learned, and focus on this one before moving to the next lesson.

⬛⬛⬛⬛⬛⬛⬛⬛ 8	Eight counted one time equals eight or 8 x 1 = 8
⬛⬛⬛⬛⬛⬛⬛⬛ 16	Eight counted two times equals sixteen or 8 x 2 = 16
⬛⬛⬛⬛⬛⬛⬛⬛ 24	Eight counted three times equals twenty-four or 8 x 3 = 24
⬛⬛⬛⬛⬛⬛⬛⬛ 32	Eight counted four times equals thirty-two or 8 x 4 = 32
⬛⬛⬛⬛⬛⬛⬛⬛ 40	Eight counted five times equals forty or 8 x 5 = 40
⬛⬛⬛⬛⬛⬛⬛⬛ 48	Eight counted six times equals forty-eight or 8 x 6 = 48

0	8	16	24	32	40	48	56	64	72	80
8×0	8×1	8×2	8×3	8×4	8×5	8×6	8×7	8×8	8×9	8×10

⬆ 8 counted 3 times

⬆ 8 counted 8 times"

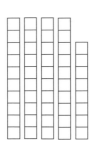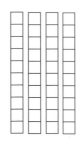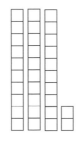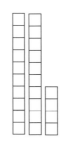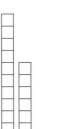

8·6=48   8·5=40   8·4=32   8·3=24   8·2=16   8·1=8

Another way to show this is on a number chart. Here we circle all of the eight facts or multiples of eight.

```
 0    1    2    3    4    5    6    7   (8)   9
10   11   12   13   14   15  (16)  17   18   19
20   21   22   23  (24)  25   26   27   28   29
30   31  (32)  33   34   35   36   37   38   39
(40) 41   42   43   44   45   46   47  (48)  49
50   51   52   53   54   55  (56)  57   58   59
60   61   62   63  (64)  65   66   67   68   69
70   71  (72)  73   74   75   76   77   78   79
(80)
```

Whenever illustrating with the blocks, also write it and say it as you build.

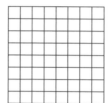

8 counted eight times is the same as 64, or
8 times 8 equals 64, or
8 over and 8 up is 64.

0 x 0	0 x 1	0 x 2	0 x 3	0 x 4	0 x 5	0 x 6	0 x 7	0 x 8	0 x 9	0 x 10
1 x 0	1 x 1	1 x 2	1 x 3	1 x 4	1 x 5	1 x 6	1 x 7	1 x 8	1 x 9	1 x 10
2 x 0	2 x 1	2 x 2	2 x 3	2 x 4	2 x 5	2 x 6	2 x 7	2 x 8	2 x 9	2 x 10
3 x 0	3 x 1	3 x 2	3 x 3	3 x 4	3 x 5	3 x 6	3 x 7	3 x 8	3 x 9	3 x 10
4 x 0	4 x 1	4 x 2	4 x 3	4 x 4	4 x 5	4 x 6	4 x 7	4 x 8	4 x 9	4 x 10
5 x 0	5 x 1	5 x 2	5 x 3	5 x 4	5 x 5	5 x 6	5 x 7	5 x 8	5 x 9	5 x 10
6 x 0	6 x 1	6 x 2	6 x 3	6 x 4	6 x 5	6 x 6	6 x 7	6 x 8	6 x 9	6 x 10
7 x 0	7 x 1	7 x 2	7 x 3	7 x 4	7 x 5	7 x 6	7 x 7	7 x 8	7 x 9	7 x 10
8 x 0	8 x 1	8 x 2	8 x 3	8 x 4	8 x 5	8 x 6	8 x 7	8 x 8	8 x 9	8 x 10	←
9 x 0	9 x 1	9 x 2	9 x 3	9 x 4	9 x 5	9 x 6	9 x 7	9 x 8	9 x 9	9 x 10
10 x 0	10 x 1	10 x 2	10 x 3	10 x 4	10 x 5	10 x 6	10 x 7	10 x 8	10 x 9	10 x 10

A stop sign is shaped like an octagon.

Sides of an Octagon - A common application of multiplying by eight is the sides of an octagon.

Example 1

How many sides are on eight stop signs?

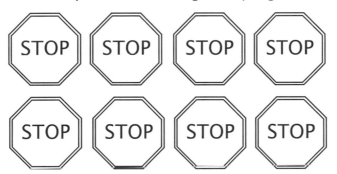

8 x 8 = 64 sides

Example 2

How many sides are on seven stop signs?

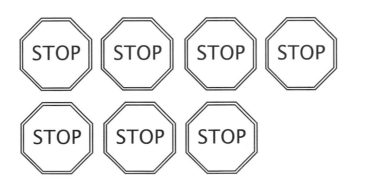

8 x 7 = 56 sides

Multiple Digit Multiplication
Place Value Notation, Distributive Property

Place-value notation is simply writing out the numbers and separating the place values. It follows the format of the blocks. For example, 123 is written as 100 + 20 + 3. This notation reinforces place value. Because it represents the blocks so well, we will use place-value notation to work each example, as well as using the manipulatives (whenever possible) and regular notation.

Remember that the *factors* are the outside dimensions of the rectangle, and the *product* is the area of the rectangle. Notice in example 1 on the next page that when building the rectangles, the up factor is on the line with the multiplication symbol, and the over factor is on the top line. Switching these factors will still produce the same answer, but it won't correspond to the picture.

Another skill presented here for the first time is the ***distributive property***. When we multiply 23 by 2 in example 1, we are multiplying the 2 times the 20 and 2 times the 3. The 2 is "distributed" among both components or place values present in 23. Verbalize this as, "Twenty-three counted two times is the same as twenty counted two times plus three counted two times." Figure 1 shows 23 multiplied by 2 linearly. While you may multiply this way, generally in this book, we will be multiplying vertically as in example 1 on the next page.

Figure 1

Distributive Property $\quad 2(23) = 2(20 + 3) = 2 \cdot 20 + 2 \cdot 3 = 40 + 6 = 46$

Example 1

$$23 \rightarrow 20+3$$
$$\underline{\times\ 2} \uparrow \underline{\times\ 2}$$
$$46 \qquad 40+6$$

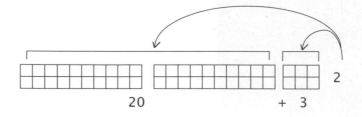

20   + 3   2

Notice that in figure 1 and example 1, we first multiply the 2 times the units (3), and then the tens (20).

Example 2

$$21 \rightarrow 20+1$$
$$\underline{\times\ 3} \uparrow \underline{\times\ 3}$$
$$63 \qquad 60+3$$

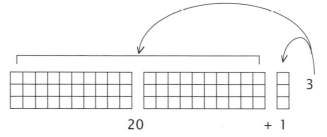

20   + 1   3

The next examples are too large to show with the blocks, but as we use the same techniques as in the first two examples, we will be fine.

Example 3

$$231 \rightarrow 200+30+1$$
$$\underline{\times\quad 3} \uparrow \underline{\times\qquad\quad 3}$$
$$693 \qquad 600+90+3$$

$$3(231) = 3(200 + 30 + 1) = 3 \cdot 200 + 3 \cdot 30 + 3 \cdot 1 = 600 + 90 + 3 = 693$$

Example 4

$$412 \rightarrow 400+10+2$$
$$\underline{\times\quad 2} \uparrow \underline{\times\qquad\quad 2}$$
$$824 \qquad 800+20+4$$

$$2(412) = 2(400 + 10 + 2) = 2 \cdot 400 + 2 \cdot 10 + 2 \cdot 2 = 800 + 20 + 4 = 824$$

When doing word problems, the student can turn a piece of notebook paper sideways and use the lines to keep his or her numbers lined up. This is especially useful when doing larger problems.

Rounding to 10, 100, and 1,000
Estimation

Most of this lesson should be review, as we have covered this material in the book preceding this one.

If rounding is new, take your time and thoroughly digest it. Rounding to 10 is used in estimating as we multiply. When you round a number to the nearest multiple of 10, there will be a number in the tens place but only a zero in the units place. I tell the students this is why we call it rounding, because the units are going to be a "round" zero.

Let's round 38 as an example. The first skill is to find the two multiples of 10 that are nearest to 38. The lower one is 30 and the higher one is 40. Thirty-eight is between 30 and 40. If the student has trouble finding these numbers, begin by placing your finger over the 8 in the units place, and all you have is a 3 in the tens place, which is 30. Then add one more to the tens to find the 40. I often write the numbers 30 and 40 above the number 38 on both sides as in figure 1.

Figure 1

         30   40
             38

The next skill is find out whether 38 is closer to 30 or 40. Let's go through all the numbers as in figure 2. It is obvious that 31, 32, 33, and 34 are closer to 30; and 36, 37, 38, and 39 are closer to 40, but 35 is right in the middle. Somebody decided that it goes to 40, so that is the reason for our rule. When rounding to tens, look at the units place. If the units are 0, 1, 2, 3, or 4, the digit in the tens place remains unchanged. If the units are 5, 6, 7, 8, or 9, the digit in the tens place increases by one. See figure 2.

Figure 2

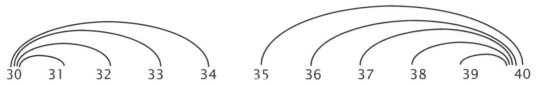

Another strategy I use is to put 0, 1, 2, 3, and 4 inside a circle to represent zero, because if these numbers are in the units place, they add nothing to the tens place. They are rounded to the lower number (30 in the example). Then I put 5, 6, 7, 8, and 9 inside a thin rectangle to represent one, because if these numbers are in the units place, they add one to the tens place and are rounded to the higher number (40 in the example).

Figure 3

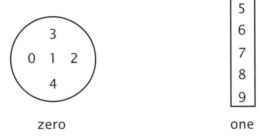

zero   one

Example 1

Round 43 to the nearest tens place.

40   50   1. Find the multiples of 10 nearest to 43.
     43

40 ↙ 50   2. We know that 3 goes to the lower number, 40.
   43

(40) ↙ 50   3. Or recall that 3 is in the circle, or 0, so nothing
    43   is added to the smaller number, 40.

When rounding to hundreds, look only at the digit in the tens place to determine whether to stay the same or increase by one. The same rules apply to hundreds as to tens. If the digit in the tens place is 0, 1, 2, 3, or 4, the number in the hundreds place remains unchanged. If the digit in the tens place is a 5, 6, 7, 8, or 9, then the number in the hundreds place increases by 1. See example 2.

Example 2

Round 547 to the nearest hundreds place.

500   600   1. Find the multiples of 100 nearest to 547.
      547

500   600   2. We know that 4 goes to the lower number, 500.
    547

(500)   600   3. Or recall that 4 is in the circle, or 0, so nothing
    547   is added to the smaller number, 500.

When rounding to thousands, consider only the number immediately to the right of the thousands place (the hundreds) to determine whether to stay the same or increase by one. See example 3.

Example 3

Round 8,719 to the nearest thousands place.

8,000   9,000   1. Find the nearest multiples of 1,000.
      8,719

8,000   9,000   2. We know that 7 goes to the higher number, 9,000.
      8,719

8,000   (9,000)   3. Or recall that 7 is in the rectangle, or 1, so
      8,719   1 is added to the 8, and the answer is 9,000.

Estimation - Now that we know how to round numbers, we can apply this skill to find the approximate answer for a multiplication problem. In example 4, we are just going to round the large factor. Later, when both factors have multiple digits, we will round both of them.

Example 4

Estimate the answer.

$$
\begin{array}{r} 43 \\ \times\ 2 \\ \hline \end{array}
\qquad
\begin{array}{r} (40) \\ \times\ \ 2 \\ \hline (80) \end{array}
$$

Round the number, and put it in the parentheses.

Multiply to find the approximation.

Example 5

Estimate the answer.

$$
\begin{array}{r} 281 \\ \times\ 3 \\ \hline \end{array}
\qquad
\begin{array}{r} (300) \\ \times\ \ \ 3 \\ \hline (900) \end{array}
$$

Round the number to the hundreds place, and put it in the parentheses.

Multiply to find the approximation.

Example 6

Estimate the answer.

$$
\begin{array}{r} 419 \\ \times\ 2 \\ \hline \end{array}
\qquad
\begin{array}{r} (400) \\ \times\ \ \ 2 \\ \hline (800) \end{array}
$$

Round the number to the hundreds place, and put it in the parentheses.

Multiply to find the approximation.

Double Digit Times Double Digit
Multiplication by 11

When multiplying two double digit numbers, you actually have four multiplication problems. Notice the four different rectangles. You begin with the units times both of the top, or over, factors, and then you multiply the tens times the top factors. The four smaller multiplication problems, or *partial products*, are: 2 x 3, 2 x 10 (not 2 x 1, but two times one ten), 10 x 3, and 10 x 10. Look at the picture and the written portion to see this unfold. When using the units, tens, and hundreds blocks, this is very clear, as the colors reveal the separate products.

Example 1

$$
\begin{array}{r}
13 \rightarrow \\
\times 12 \uparrow \\
\hline
26 \\
130 \\
\hline
156
\end{array}
\qquad
\begin{array}{r}
10+3 \\
\times 10+2 \\
\hline
20+6 \\
100+30+ \\
\hline
100+50+6
\end{array}
$$

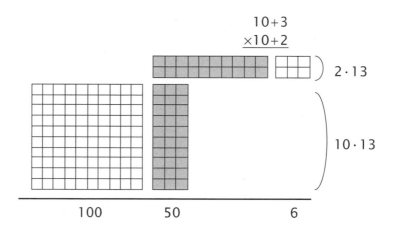

You can also see in example 1 that it is not only four smaller problems. It can also be broken down into two problems. The problem 13 x 12 can be broken down into 10 · 13 and 2 · 13. Double-digit multiplication is a conjoining of these two problems into one. You can see this in the figure on the previous page, with the factors to the right of the drawing.

$$\begin{array}{r} 13 \rightarrow \\ \times 10 \uparrow \\ \hline 130 \end{array} \qquad \begin{array}{r} 13 \rightarrow \\ \times\ 2 \uparrow \\ \hline 26 \end{array}$$

Have you ever wondered why you shift the second line of a problem of this nature? It is because of place value. After building the rectangle, ask the student, "Is every value is in its own place?" In the written portion, we wrote in the place values (thus all the zeroes). If we understand place value, we don't need to write all the zeroes but just put everything in its place. The reason for the shift to the left is to put every value in its own place, so that we can add the units to the units, the tens to the tens, and the hundreds to the hundreds.

Example 2

$$\begin{array}{r} 14 \\ \times 12 \\ \hline 28 \\ 140 \\ \hline 168 \end{array} \qquad \begin{array}{r} 10+4 \\ \times 10+2 \\ \hline 20+8 \\ 100+40 \\ \hline 100+60+8 \end{array}$$

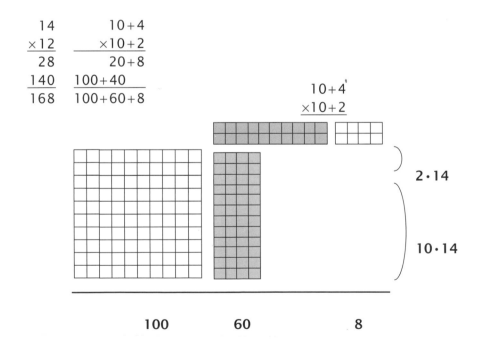

Example 3

```
 22      20+2
×13      ×10+3
 66       60+6
220    200+20
286    200+80+6
```

Example 4 is an example of a three-digit number times a two-digit number. Notice that it is just a combination of what we have been doing. Notice also how it is two problems being solved together.

Example 4

```
  212      212        212           200+10+2
 ×10   +  × 4   =    ×14   or      ×   10+4
2,120     848        848            800+40+8
                   2,120          2,000+100+20
                   2,968          2,000+900+60+8
```

Multiplication by 11 - When you multiply by 11, there is an interesting phenomenon that occurs. I call it split 'em and add 'em. This works only if the number being multiplied by 11 has only two digits and those digits don't add up to more than 9.

Examples 5 and 6 qualify. The pattern is to split 'em (the digits in 23) to 2 _ 3, and then add 'em, 2 + 3 = 5, for the middle term. You get 253. Looking at the examples, do you see where this originates? When you multiply by 11, the digits in the partial product are the same. This is not critical to learn, but it's a fun observation.

Example 5

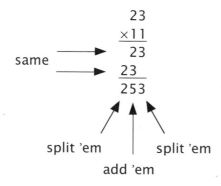

Example 6

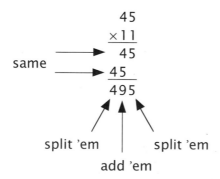

Double Digit with Regrouping
Mental Math

The way that I present this topic is not the traditional approach. It has distinct advantages and is revealed by the manipulatives, but it is different. Consider this method at least as an alternative for regrouping in a multiplication problem.

Before double digit times double digit with regrouping, let's look at a single-digit number times a double-digit number with regrouping

Example 1

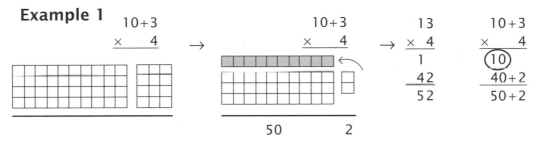

After multiplying you have 12 in the units place, and need to regroup. The arrow and the shading indicate the 10 moving to the tens place. Instead of carrying the 1 up above the 1 in the tens place, put it in the tens place below the line. Then add as usual.

Example 2

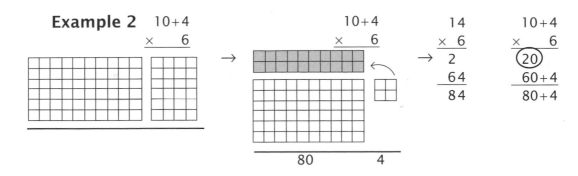

After multiplying in example 2, you have 24 in the units place and need to regroup. Instead of carrying the 2 up above the 2 in the tens place, put it in the tens place below the line. Then add.

Example 3

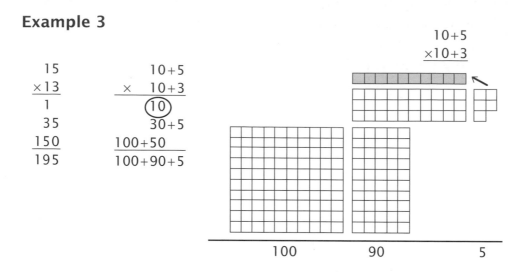

$$\begin{array}{r} 15 \\ \times 13 \\ \hline 1 \\ 35 \\ 150 \\ \hline 195 \end{array} \qquad \begin{array}{r} 10+5 \\ \times \quad 10+3 \\ \hline \textcircled{10} \\ 30+5 \\ 100+50 \\ \hline 100+90+5 \end{array}$$

Notice that after multiplying, you have 15 in the units place and need to regroup. Instead of carrying the 1 up above the 1 in the tens place, put it in the tens place below the line. Then add as usual. To help keep the place values properly aligned, here are two tips. First, either use graph paper or lined paper turned sideways to keep the digits in the proper place. Second, when you add you can circle any digits that are "carried." This helps remove potential confusion.

Example 4

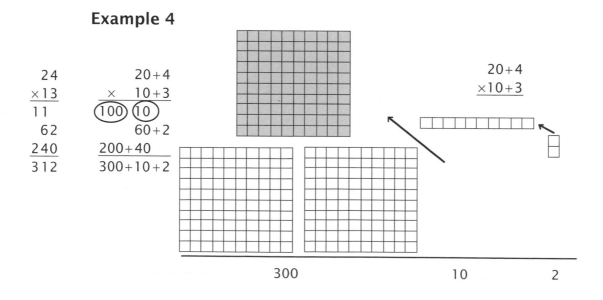

$$\begin{array}{r} 24 \\ \times 13 \\ \hline 11 \\ 62 \\ 240 \\ \hline 312 \end{array} \qquad \begin{array}{r} 20+4 \\ \times \quad 10+3 \\ \hline \textcircled{100}\ \textcircled{10} \\ 60+2 \\ 200+40 \\ \hline 300+10+2 \end{array}$$

In example 4, we regroup the initial problems, and the arrow in the units place shows this. When adding the tens after the multiplying is done, we regroup 10 tens for 1 hundred. The hundred is shaded as the ten was in previous examples.

When using the traditional way, after you multiply the units, you place the 1 above the tens place as shown in example 5. Then you multiply 3 x 1 to get 3, and add the 1 to make it 4. There are two potential problems that this creates. First, many students never understand why you mix the operations and multiply 3 x 1, and then *add* the other 1. Second, in a problem with a lot of regrouping, you can develop quite a few numbers above the top factor, and then the student adds the wrong numbers.

With my method, you do all the multiplying first, and then do all the adding below the line. In some cases, it is not only clearer, but quicker. Some older students who have already learned the traditional way won't like to switch, so show the method to them and make it optional. Both ways work.

Example 5

```
    15            10+5
  ×13          ×   10+3
    1             10
    35            30+5
  150          100+50
  195          100+90+5
```

Here are some more examples to study. Make your own decision about what method you prefer your student to use. Unfortunately, these problems are too large to show with the blocks.

Example 6

traditional way

```
   37           30+7              2
 ×14          ×  10+4            37
   12         ⟨100⟩ ⟨20⟩       ×14
  ₁28         ⟨100⟩ 20+8          1
 370          300+70            128
 518          500+10+8         370
                               518
```

Example 7

traditional way

```
    28          20+8              15
  ×27       ×   20+7              28
    1         (100)             ×27
   15       (100)(50)             1
  146       (100) 40+6          196
  460        400+60            560
  756        700+50+6          756
```

Mental Math - These problems can be used to keep the facts alive in the memory and to develop mental math skills. Try a few at a time, saying the problem slowly so that the student comprehends. The purpose is to stretch but not discourage. You decide where that line is! Lessons 27 and 30 have more of these.

Example 8

"Two times three, times one, equals what number?"

The student thinks, "2 x 3 = 6, and 6 x 1 = 6." At first, go slowly enough for him to verbalize the intermediate step. As skills increase, the student should be able to say the just the answer.

1. Two times three, times eight, equals what number? (48)

2. One times seven, times five, equals what number? (35)

3. Four times two, times nine, equals what number? (72)

4. Three times three, times three, equals what number? (27)

5. Five times one, times six, equals what number? (30)

6. Two times two, times eight, equals what number? (32)

7. Three times two, times two, equals what number? (12)

8. One times nine, times seven, equals what number? (63)

9. Seven times two, times zero, equals what number? (0)

10. Five times four, times one, equals what number? (20)

Multiple Digit Multiplication, Regrouping

This is not much different from the previous lesson, except now we are multiplying three digits times two digits and regrouping. In example 1, we see that it is still a combination of two problems since the factor doing the multiplying has two digits. This factor (17 in example 1) is called the *multiplier*. The top factor (245 in example 1) is referred to as the *multiplicand*. Of course, the answer is the *product*. The estimates are in the parentheses.

Also in this lesson, we are using *ten thousands*. See figure 1. Do you also see that within the "thousands" there are 1, 10, and 100, just as within the units? As you move from right to left there is a progression of multiplying by a factor of 10. Ten times 1 is 10, 10 times 10 is 100, and 10 times 100 is 1,000. This continues in the thousands. Ten times 1,000 is 10,000, and 10 times 10,000 is 100,000. A comma separates the units and the thousands. I like to think of this comma as representing the word "thousand" when reading a number. In example 2, the estimated answer 18,000 is read as "eighteen thousand."

Figure 1

Example 1

```
  245          245            245           (2 0 0)        (2⎛
× 7       +  × 10     =     × 17          × (2 0)        × (2⎝
¹23          2,450         1 ¹23          (4,0 0 0)      (4,0 0 0)
1,485                      1,485
1,715                      2,450
                           4,165
```

Example 2

```
  163          163            163           (2 0 0)        (2⎛
× 9       +  × 80     =     × 89          × (9 0)        × (9⎝
 52          ¹4 2            ¹52          (18,0 0 0)     (18,0 0 0)
¹947         8,840          ² 947
1,467        13,040         14 2
                            8,840
                            14,5 07
```

Example 3

```
  412          412            412           (4 0 0)        (4⎛
×  7      +  × 50     =     × 57          × (6 0)        × (6⎝
 2  1                       1            (24,0 0 0)     (24,0 0 0)
  874          1          2   1
 2,884       20,500        1  874
             20,600        1
                          2¹0,500
                          2 3,484
```

Example 4

```
  208          208            2 08          (2 0 0)        (2⎛
×  3      +  × 60     =     × 63          × (6 0)        × (6⎝
  2          4              1 2          (12,0 0 0)     (12,0 0 0)
 604         12,080         1 6 04
 624         12,480        12,⁴0 80
                           13, 104
```

Finding Factors, 25¢ = 1 Quarter

Factoring is the opposite of multiplying. To multiply you are given two factors, and you build a rectangle to find the product. To factor, you are given the area, or product, and you build a rectangle to find the factors. Let's work through the first example . Begin by asking, "What are the factors of six?" Then take six green blocks (or three orange blocks), and see how many different rectangles you can make with those blocks. In this case, you can build two rectangles, one by six and two by three. So the factors of six are 1 x 6 and 2 x 3.

Example 1

Find the factors of 6.

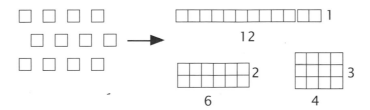

The factors are 1 x 6 and 2 x 3.

Example 2
Find the factors of 12.

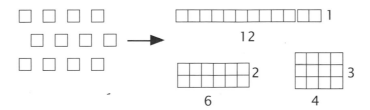

The factors are 1 x 12, 2 x 6, and 3 x 4.

Example 3

Find the factors of 15.

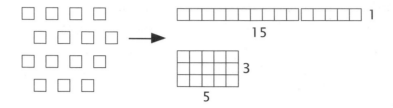

The factors are 1 x 15 and 3 x 5.

Example 4

Find the factors of 24.

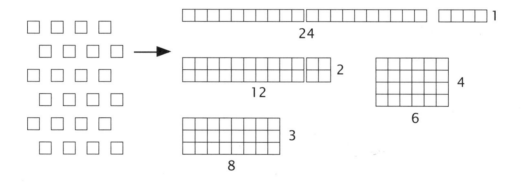

The factors are 1 x 24, 2 x 12, 3 x 8, and 4 x 6.

When factoring, use what you know about the multiplication facts to help you find potential factors. If the number is even, then you know it will have a factor of two. If the digits add up to three or a multiple of three, then you know that three is a factor. If a number ends in zero, then we know that it is a multiple of 10. If a number ends in a five or a zero, then five is a factor. And if the digits add up to nine or a multiple of nine, then nine will also be a factor. Knowing your multiplication facts and recognizing patterns in multiplication is the best preparation for factoring.

25¢ = 1 Quarter of a Dollar - In lesson 16 we learned that there are four quarters in a dollar, and that there are 25¢ in a quarter. Now that we know how to regroup while multiplying double-digit numbers, we can solve how many pennies are in any number of quarters. See examples 5, 6, 7, and 8.

25¢ = "twenty-five cents" = <image: quarter> = <image: blocks>

Example 5

How many pennies are in three quarters?

<image: three quarters>

$$
\begin{array}{r}
2\,5 \\
\times\ 3 \\
\hline
1 \\
65 \\
\hline
75
\end{array}
\qquad
\begin{array}{r}
20+5 \\
\times\ \ 3 \\
\hline
1 \\
60+5 \\
\hline
70+5
\end{array}
$$

Example 6

How many pennies are in seven quarters?

$$
\begin{array}{r}
25 \\
\times\ 7 \\
\hline
3 \\
145 \\
\hline
175
\end{array}
\qquad
\begin{array}{r}
20+5 \\
\times\ \ 7 \\
\hline
30 \\
100+40+5 \\
\hline
100+70+5
\end{array}
$$

Example 7

How many pennies are in 11 quarters?
(Remember split 'em and add 'em?)

```
    25              20+5
   ×11           ×10+1
    25              20+5
   25           200+50
   275          200+70+5
```

Example 8

How many pennies are in 25 quarters?

```
    25              20+5
   ×25           ×20+5
     2              20
   105           100+ 0+5
     1           100
   40            400+ 0
   625          600+20+5
```

Place Value Through Millions
16 ounces = 1 Pound

Understanding place value helps you know where to place the digits in multiple-digit multiplication. For some, this will be review. The beginning value we call units. This is represented by the small green half-inch cube. The next largest place value is the tens place, shown with the blue 10 bar. This is 10 times as large as a unit. The next value we come to, as we move to the left, is the hundreds, shown by the large red block. It is 10 times as large as a 10 bar. Notice that as you move to the left, each value is 10 times as large as the preceding value. See figure 1.

Figure 1

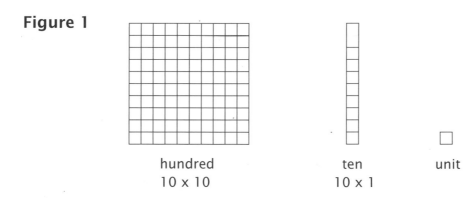

hundred
10 x 10

ten
10 x 1

unit

When you name a number such as 247, the 2 tells you how many hundreds, the 4 indicates how many tens, and the 7 shows how many units. We read 247 as "two hundred for-ty (ty means 10) seven." The 2, 4, and 7 are digits that tell us how many. The hundreds, tens, and units tell us what kind or what value. Where the digit is written or what *place* it occupies, tells us what *value* it is. Notice that as the values progress from right to left, they keep increasing by a factor of 10. That is because we are operating in the base 10 system or the decimal system.

The next place value is the thousands place. It is 10 times 100. You can build 1,000 by stacking 10 hundreds squares and making a cube. You can also show 1,000 by making a rectangle that is 10 by100 out of the cube, as in figure 2. You will see that I used a much smaller scale to be able to show this. Also in figure 2 is 10,000. If you stick to rectangles, can you imagine what 1,000,000 would look like? It would be a rectangle 100 by1,000. The factors are inside the rectangles.

Figure 2

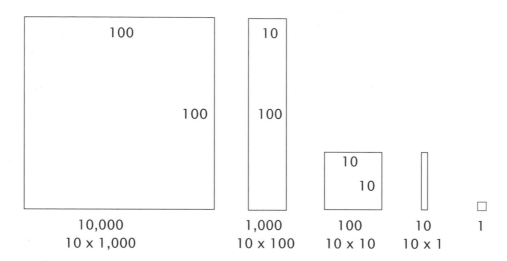

Notice the progression of multiplying by a factor of 10 as you move from right to left. Ten times 1,000 is 10,000 and 10 times 10,000 is 100,000. Ten times 100,000 is 1,000,000 (one million).

Do you also see that within the "thousands," there are 1, 10, and 100, just as in the units that are shown in figure 1? The same is true with the millions. There is 1 million, 10 million, and 100 million. The commas separate the units, thousands, and millions. See figure 3.

Figure 3

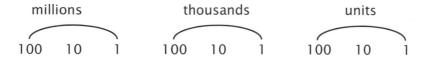

When saying these larger numbers, I like to think of the commas as having names. The first comma from the right is named "thousand," and the second from the right is "million." See example 1.

Example 1

Say 123,456,789.

"123 million, 456 thousand, 789" or
"one hundred twenty-three million, four hundred fifty-six thousand, seven hundred eighty-nine."

Notice that you never say "and" when reading a large number. This is reserved for the decimal point in a succeeding book. Do you see that you never read a number larger than hundreds between the commas? This is because there are only three places between the commas. Practice saying and writing larger numbers.

Place-Value Notation - This is a way of writing numbers that emphasizes the place value. See example 2.

Example 2

Write 8,543,971 with place-value notation.

8,000,000 + 500,000 + 40,000 + 3,000 + 900 + 70 + 1

Weight - 16 Ounces = 1 Pound - Begin by holding up three Math-U-See blue 10 bars. The weight of these three added together is really close to one ounce. Sixteen ounces is the same as one pound. Four red 100 bars plus five blue 10 bars is approximately one pound. This will give the student a feel for one ounce and one pound. The abbreviation for ounce is *oz* and the abbreviation for pound is *lb*.

Example 3

How many ounces are in six pounds?

If ☐ = 1 ounce, then ☐ = 16 ounces = 1 pound.

$$
\begin{array}{r}
16 \\
\times\ 6 \\
\hline
3 \\
66 \\
\hline
96
\end{array}
\qquad
\begin{array}{r}
10+6 \\
\times\quad 6 \\
\hline
30 \\
60+6 \\
\hline
90+6
\end{array}
$$

Example 4

How many ounces are in 24 pounds?

```
     16              10+6
   ×24          ×    20+4
     2               20
    44               40+4
   1             100
   22            200+20
  384            300+80+4
```

Mental Math - Here are some more questions to read to your student. These include multiplication and addition. Remember to go slowly at first.

1. One times three, plus nine, equals what number? (12)

2. Four plus six, times five, equals what number? (50)

3. Three times two, plus three, equals what number? (9)

4. Four times eight, plus one, equals what number? (33)

5. Six plus three, times eight, equals what number? (72)

6. Three times three, plus eight, plus one, equals what number? (18)

7. Five plus two, times seven, plus zero, equals what number? (49)

8. Two plus eight, times four, plus five, equals what number? (45)

9. Eight times two, plus one, plus one, equals what number? (18)

10. Two times four, plus one, times five, equals what number? (45)

LESSON 28

More Multiple Digit Multiplication

There are no new concepts in this lesson, just larger numbers. Take your time and be careful of the place value, and you will do fine. Estimation will help you recognize errors, especially when you use a calculator. If you want to use one to check your answers, this is a good time to do so. But don't let a machine take the place of good thinking!

Example 1

$$
\begin{array}{r}
392 \\
\times 147 \\
\end{array}
\quad = \quad
\begin{array}{r}
392 \\
\times 100 \\
\hline
39{,}200 \\
\end{array}
\quad + \quad
\begin{array}{r}
392 \\
\times 40 \\
\hline
3 \\
12{,}680 \\
\hline
15{,}680 \\
\end{array}
\quad + \quad
\begin{array}{r}
392 \\
\times 7 \\
\hline
61 \\
2{,}134 \\
\hline
2{,}744 \\
\end{array}
$$

$$
\begin{array}{r}
392 \\
\times 147 \\
\hline
{}^{1}61 \\
{}^{1}2{,}134 \\
3 \\
{}^{1}12{,}680 \\
39{,}200 \\
\hline
57{,}624 \\
\end{array}
$$

$$
\begin{array}{r}
(4 \\
\times (1 \\
\hline
(40{,}000) \\
\end{array}
\qquad
\begin{array}{r}
(400) \\
\times (100) \\
\hline
(40{,}000) \\
\end{array}
$$

You might wonder why the answer is not very close to the estimate. Notice that 147 is rounded appropriately to 100. But it is only three away from being 150, which would be rounded to 200. So we can expect the answer to be between 100 times 400 (40,000) and 200 times 400 (80,000), which is 60,000. And that is what we find. 57,624 is almost 60,000.

Example 2

$$
\begin{array}{r}
655 \\
\times 708 \\
\end{array}
$$

$$
=
\begin{array}{r}
655 \\
\times 700 \\
\hline
3\,3 \\
425,500 \\
\hline
458,500 \\
\end{array}
\;+\;
\begin{array}{r}
655 \\
\times\ \ \ 0 \\
\hline
0 \\
\end{array}
\;+\;
\begin{array}{r}
655 \\
\times\ \ \ 8 \\
\hline
44 \\
{}^{l}4,800 \\
\hline
5,240 \\
\end{array}
$$

$$
\begin{array}{r}
655 \\
\times\ 708 \\
\hline
44 \\
{}^{l}4,800 \\
{}^{l}3\ 3 \\
42\ 5,500 \\
\hline
46\ 3,740 \\
\end{array}
$$

$$
\begin{array}{r}
(7\ \\
\times\ (7\ \\
\hline
(490,0\,00) \\
\end{array}
\qquad
\begin{array}{r}
(700) \\
\times\ (700) \\
\hline
(490,0\,00) \\
\end{array}
$$

Notice that we don't have a separate line for 0 times 655. Since we understand place value, we know that when multiplying by the hundreds, we begin with the hundreds place. In this case, that is the 5 in 425,500. Or if we write the two zeros to show that we are in the hundreds place, we are all set as well.

Example 3

$$
\begin{array}{r}
4,382 \\
\times 961 \\
\end{array}
$$

$$
=
\begin{array}{r}
4,382 \\
\times 900 \\
\hline
{}^{l}27\ 1 \\
3,6\,72,800 \\
\hline
3,9\,43,800 \\
\end{array}
\;+\;
\begin{array}{r}
4,382 \\
\times\ \ 60 \\
\hline
{}^{l}1\,4\ 1 \\
248,820 \\
\hline
262,920 \\
\end{array}
\;+\;
\begin{array}{r}
4,382 \\
\times\ \ \ 1 \\
\hline
4,382 \\
\end{array}
$$

$$
\begin{array}{r}
4,3\,82 \\
\times\ 9\,61 \\
\hline
{}^{2}4,{}^{l}382 \\
{}^{2}1\,4\ 1 \\
{}^{2}24\,8,8\,20 \\
2\,7\ 1 \\
{}^{l}3\,6\,7\,2,8\,00 \\
\hline
4,2\,1\,1,10\,2 \\
\end{array}
$$

$$
\begin{array}{r}
(4,\ \ \ ) \\
\times\ (1,\ \ \ ) \\
\hline
(4,00\,0,000) \\
\end{array}
\qquad
\begin{array}{r}
(4,000) \\
\times\ (1,000) \\
\hline
(4,00\,0,000) \\
\end{array}
$$

Prime and Composite Numbers
Multiply by 12

In lesson 26, we learned how to find the factors of a number by building a rectangle and reading its dimensions. In all of the examples, we could build at least two different rectangles with different dimensions, which means there were at least two different sets of factors. Remember that one by six and six by one are the same rectangle. One may be vertical and the other horizontal, but they have the same factors.

A number like 6 or 12, from which you can build more than one rectangle, is a *composite number.* All the numbers given in lesson 26 were composite numbers. The numbers between 2 and 24 that are composite are: 4, 6, 8, 9, 10, 12, 14, 15, 16, 18, 20, 21, 22, and 24.

A number that has only one set of factors, or from which you can build only one rectangle, is called a *prime number.* The definition of a prime number is "any number that has only the factors of one and itself" or "can be divided evenly only by one and itself." (The number one is not considered a prime number.) The prime numbers between 2 and 24 are: 2, 3, 5, 7, 11, 13, 17, 19, and 23. Ask the student to build some of them, and he or she will find that there is only one rectangle to be built and thus only one set of factors. Example 1 illustrates a prime number and example 2 illustrates a composite number.

Example 1

Find all possible factors of 13, and tell whether it is prime or composite.

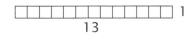

13 has only one set of factors, 1 x 13, so it is prime.

Example 2

Find all possible factors of 20, and tell whether it is prime or composite.

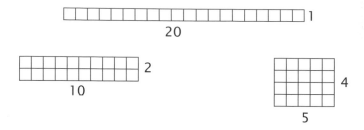

Twenty has three sets of factors, 1 x 20, 2 x 10, and 4 x 5, so it is composite.

Multiply by 12 - Some teachers present multiplying by 12 (the 12 facts) as a set of facts to be memorized. This is a good idea since we use multiples of 12 in many areas of life. There are 12 items in a dozen, 12 inches in a foot, 12 months in a year, and 12 hours on a clock face. Twelve is also a double-digit number, and with a few strategies we can learn how to multiply by it fairly easily. Memorization is a good thing as well. Practice 1·12 through 12·12 until you can do the 12 facts quickly.

The number 12 is 10 + 2. So when you multiply 4 x 12, think 4(10 + 2), and use the distributive property: 4 x 12 = 4(10 + 2) = 4 x 10 + 4 x 2 = 40 + 8 = 48. Take the multiplier times the tens, and then times the units, and add the answers. You are breaking a hard problem into two easier problems.

Example 3

Solve 8 x 12. 8 x 12 = 8(10 + 2) = 8 · 10 + 8 · 2 = 80 + 16 = 96 So 8 x 12 = 96

Example 4

Solve 6 x 12. 6 x 12 = 6(10 + 2) = 6 · 10 + 6 · 2 = 60 + 12 = 72 So 6 x 12 = 72

Example 5

Solve 11 x 12. I picture a rectangle over 11 and up 12. This is 132. See the rectangle to the right.

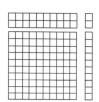

5,280 Feet = 1 Mile; 2000 Pounds = 1 Ton

Find a straight stretch of road in your area that you can measure with the odometer in your car to show how far a mile is. Or look for mile markers, which are on most interstates and turnpikes. If there is a track around a football field nearby, find out how long it is, and define a mile in terms of how many laps around it. Most tracks are either one-fourth or one-third of a mile.

There are 5,280 feet in one mile. Since this is true, how many feet are there in six miles? See example 1.

Example 1

How many feet in six miles?

5,280 x 6 = 31,680 ft

```
  5,280        (5,
 ×    6       × (6  )
   1 4        (30,000)
 30,280
 31,680
```

Example 2

How many feet in 25 miles?

5,280 x 25 = 132,000 ft

```
   5,280        (5,
 ×    25       × (3
   1 1         (150,000)
   1 4
 25,000
     1
 104,600
 132,000
```

Weight - 2,000 Pounds = 1 Ton - One thousand is 10 x 10 x 10, or 10 x 100, or 10 red hundred blocks. Two thousand is double that or 20 red hundred blocks. If one green unit is one pound, then 20 hundred blocks is 2,000 pounds (lb).

If ☐ = 1 pound, then [grid] x 20 = 2,000 pounds = 1 ton

Example 3

How many pounds in four tons? 2,000 x 4 = 8,000 pounds

$$
\begin{array}{r} 2, \\ \times \quad 4 \\ \hline 8,000 \end{array}
\qquad
\begin{array}{r} 2,000 \\ \times \quad 4 \\ \hline 8,000 \end{array}
$$

Example 4

How many pounds in 27 tons? 2,000 x 27 = 54,000 lb

$$
\begin{array}{r} 2,000 \\ \times \quad 27 \\ \hline 14,000 \\ 40,000 \\ \hline 54,000 \end{array}
\qquad
\begin{array}{r} (2,0 \\ \times \quad (3 \\ \hline (6\,0,0\,0\,0) \end{array}
$$

Mental Math - Here are some more questions to read to your student. These include multiplication, addition and subtraction.

1. Six minus three, plus five, times one, equals what number? (8)

2. Four times three, minus eight, plus seven, equals what number? (11)

3. Seventeen minus nine, plus one, times six, equals what number? (54)

4. Two times two, plus seven, minus one, equals what number? (10)

5. Seven times three, minus one, plus six, equals what number? (26)

6. One times ten, plus eight, minus nine, equals what number? (9)

7. Twelve minus seven, plus four, times three, equals what number? (27)

8. Six times three, plus two, minus ten, equals what number? (10)

9. Five plus zero, times eight, minus twenty, equals what number? (20)

10. Fifteen minus eight, plus three, times ten, equals what number? (100)

APPENDIX

Nineovers for Multiplication

There is no check for multiplying large numbers, other than long division. While the technique taught here is not foolproof, it is a quick and effective way to test the answer to a multiplication problem. A similar skill is taught in detail in the appendix to *Beta,* showing how to do this with addition. Examples 1 through 4 illustrate striking nines with addition.

Until the advent of calculators, this skill was widely known and used by accountants and bookkeepers. There are various names for this method of checking; some call it casting out nines, and others refer to it as striking nines. I have coined the name "nineovers" since it is based on dividing by nine and counting what is left over or the remainder. Since the student hasn't learned how to divide by nine yet, the explanation is for the teacher and can be shared with the student at the appropriate time. We'll begin with the concept, or the why, of nineovers.

If you divide 27 by 9, the answer is 3, with a remainder of 0. We could have predicted this remainder by adding up the digits of 27. (2 + 7 = 9.) We learned in multiplication by nine that a characteristic shared by all the multiples of nine is that the digits add up to nine or a multiple of nine. Notice this in 18, 27, 36, 45, 54, etc. The digits add up to nine. In 963, the digits add up to 18, which is a multiple of nine. If you were to divide 963 by 9, you would have an answer of 107, with a remainder of 0.

Consider the number 274. If you divide it by 9, the answer is 30 with a remainder of 4. (2 + 7 = 9, with 4 left over.) You can predict the remainders, or "nineovers," of a number that is divided by nine by adding up the digits and subtracting nine. In the number 138, the remainder is 3 because 1 + 8 = 9. If you were adding 274 + 138, the remainder for the sum should be 7 (4 from 274 and

3 from 138). If the remainder were 6 or 8, then you would expect an error in your calculations and go back to rework the problem. See figure 1.

Figure 1

$$
\begin{array}{rcl}
274 & \rightarrow & 4 \\
+138 & \rightarrow & 3 \\
\hline
412 & \rightarrow & 7
\end{array}
$$

Example 1

$$
\begin{array}{l}
274 \\
+138 \\
\hline
412
\end{array}
\qquad
\begin{array}{lcl}
\cancel{2} + \cancel{7} + 4 & \rightarrow & 4 \\
\cancel{1} + 3 + \cancel{8} & \rightarrow & 3 \\
4 + 1 + 2 & \rightarrow & 7
\end{array}
$$

Strike out the combinations of nine in each addend. Then add the nineovers. 4 + 3 = 7.

Compare this answer with adding up the digits in the sum. 4 + 1 + 2 = 7. They agree.

Example 2

$$
\begin{array}{l}
274 \\
+138 \\
\hline
412
\end{array}
\qquad
\begin{array}{lcl}
2 + 7 + 4 = 13 & \rightarrow & 1 + 3 = 4 \\
1 + 3 + 8 = 12 & \rightarrow & 1 + 2 = 3 \\
4 + 1 + 2 = 7 & \rightarrow & 7
\end{array}
$$

Add the digits in the addends until the answer is a single digit. 2 + 7 + 4 = 13 and then 1 + 3 = 4.

Compare this answer with adding up the digits in the sum. 4 + 1 + 2 = 7. They agree.

Example 3

Strike out the nines in the addends.

2731	$2 + 7 + 3 + 1 \rightarrow 13 \rightarrow 1 + 3 = 4$
4251	$4 + 2 + 5 + 1 \rightarrow 12 \rightarrow 1 + 2 = 3$
+2120	$2 + 1 + 2 + 0 \rightarrow 5 \rightarrow \quad = 5$
9,102	$9 + 1 + 0 + 2 \rightarrow 12 \rightarrow 1 + 2 = 3$

Add the digits in the addends until the answer is a single digit. $4 + 3 + 5 = 12$ and then $1 + 2 = 3$. Compare this answer with adding up the digits in the sum. $1 + 2 = 3$. They agree.

When multiplying, instead of adding the remainders, you multiply them and then compare with the product. This makes sense when you consider that multiplication is fast adding of the same number. In short, when adding, add the remainders or nineovers, and when multiplying, multiply the nineovers. The next examples are all with multiplication. All the work is not shown, so we can focus on comparing the factors and the product.

Example 4

Strike out the nines in the factors. Multiply the nineovers: $2 \times 4 = 8$

254	$2 + 5 + 4 \rightarrow 2$
× 76	$7 + 6 \rightarrow 13 \rightarrow 1 + 3 = 4$
19,304	$1 + 9 + 3 + 0 + 4 \rightarrow 1 + 7 = 8$

Compare this answer with adding up the digits in the product to get 8. They agree.

Example 5

391	3 + 9̶ + 1 →	4
× 80	8 + 0 →	8
31,280	3 + 1̶ + 2 + 8̶ + 0 → 3 + 2 + 0 = 5	

Strike out the nines in the factors. Multiply the nineovers. 4 x 8 = 32 and 3 + 2 = 5.

Compare this answer with adding up the digits in the product to get 5. They agree.

Example 6

125	1 + 2 + 5 →	8
× 24	2 + 4 →	6
3,000	3 + 0 + 0 + 0 →	3

Strike out the nine in the factors Multiply the nineovers. 6 x 8 = 48, 4 + 8 = 12 and 1 + 2 = 3.

Compare this answer with adding up the digits in the product to get 3. They agree.

Student Solutions

Lesson Practice 1A
1. done
2. $3 \times 3 = 9$
3. $2 \times 6 = 12$
   $6 \times 2 = 12$
4. $3 \times 4 = 12$
   $4 \times 3 = 12$
5. $5 \times 5 = 25$
6. $2 \times 4 = 8$
   $4 \times 2 = 8$
7. 3x4 rectangle
8. 3x5 rectangle

Lesson Practice 1B
1. $2 \times 2 = 4$
2. $4 \times 6 = 24$
   $6 \times 4 = 24$
3. $2 \times 5 = 10$
   $5 \times 2 = 10$
4. $4 \times 5 = 20$
   $5 \times 4 = 20$
5. $4 \times 7 = 28$
   $7 \times 4 = 28$
6. $2 \times 3 = 6$
   $3 \times 2 = 6$
7. 3x5 rectangle
8. 7x3 rectangle

Lesson Practice 1C
1. $2 \times 7 = 14$
   $7 \times 2 = 14$
2. $3 \times 5 = 15$
   $5 \times 3 = 15$
3. $3 \times 7 = 21$
   $7 \times 3 = 21$
4. $4 \times 4 = 16$

5. $1 \times 8 = 8$
   $8 \times 1 = 8$
6. $4 \times 3 = 12$
   $3 \times 4 = 12$
7. 2x5 rectangle
8. 5x2 rectangle

Lesson Practice 1D
1. $5 \times 2 = 10$
   $2 \times 5 = 10$
2. $4 \times 7 = 28$
   $7 \times 4 = 28$
3. $3 + 1 = 4$
4. $4 + 2 = 6$
5. $2 + 8 = 10$
6. $9 + 5 = 14$
7. $8 + 6 = 14$
8. $5 + 5 = 10$
9. $7 + 6 = 13$
10. $4 + 5 = 9$
11. $10 - 1 = 9$
12. $8 - 2 = 6$
13. $15 - 9 = 6$
14. $16 - 8 = 8$
15. $8 - 4 = 4$
16. $7 - 3 = 4$
17. $9 - 6 = 3$
18. $11 - 7 = 4$

Systematic Review 1E
1. $3 \times 2 = 6$
   $2 \times 3 = 6$
2. $5 \times 3 = 15$
   $3 \times 5 = 15$
3. $6 + 1 = 7$
4. $5 + 2 = 7$

5. $2 + 3 = 5$

6. $9 + 7 = 16$

7. $8 + 4 = 12$

8. $6 + 6 = 12$

9. $5 + 6 = 11$

10. $4 + 7 = 11$

11. $8 - 1 = 7$

12. $5 - 2 = 3$

13. $18 - 9 = 9$

14. $14 - 7 = 7$

15. $15 - 8 = 7$

16. $8 - 5 = 3$

17. $9 - 8 = 1$

18. $12 - 4 = 8$

Systematic Review 1F

1. $8 \times 1 = 8$
   $1 \times 8 = 8$

2. $3 \times 3 = 9$

3. $7 + 1 = 8$

4. $6 + 2 = 8$

5. $2 + 9 = 11$

6. $9 + 4 = 13$

7. $8 + 3 = 11$

8. $4 + 4 = 8$

9. $7 + 8 = 15$

10. $5 + 7 = 12$

11. $3 - 1 = 2$

12. $9 - 2 = 7$

13. $16 - 9 = 7$

14. $13 - 8 = 5$

15. $12 - 6 = 6$

16. $7 - 4 = 3$

17. $13 - 7 = 6$

18. $12 - 3 = 9$

Lesson Practice 2A

1. $1 \times 0 = 0$

2. $0 \times 3 = 0$

3. $1 \times 0 = 0$

4. $0 \times 6 = 0$

5. $9 \times 0 = 0$

6. $0 \times 2 = 0$

7. $0 \times 5 = 0$

8. $8 \times 0 = 0$

9. $7 \times 0 = 0$

10. $1 \times 1 = 1$

11. $8 \times 1 = 8$

12. $1 \times 2 = 2$

13. $3 \times 1 = 3$

14. $1 \times 5 = 5$

15. $7 \times 1 = 7$

16. $1 \times 4 = 4$

17. $9 \times 1 = 9$

18. $1 \times 6 = 6$

19. $1 \times 0 = 0$

20. $0 \times 5 = 0$

21. $4 \times 1 = 4$
    $1 \times 4 = 4$

22. $0 + 0 + 0 + 0 + 0 + 0 = 0$

23. $8 \times 1 = 8$

24. $3 \times 1 = 3$ flowers

25. $9 \times 1 = 9$ bookmarks

Lesson Practice 2B

1. $5 \times 0 = 0$

2. $0 \times 1 = 0$

3. $6 \times 1 = 6$

4. $1 \times 9 = 9$

5. $1 \times 4 = 4$

6. $7 \times 1 = 7$

7. $1 \times 5 = 5$

8. $3 \times 1 = 3$

9. $1 \times 2 = 2$

10. $8 \times 1 = 8$

11. $1 \times 1 = 1$

12. $7 \times 0 = 0$

13. $8 \times 0 = 0$

14. $0 \times 5 = 0$

15. $0 \times 2 = 0$

16. $0 \times 9 = 0$

17. $0 \times 1 = 0$

18. $3 \times 0 = 0$
19. $0 \times 4 = 0$
20. $6 \times 0 = 0$
21. $5 \times 1 = 5$
    $1 \times 5 = 5$
22. $0 = 0$
23. $10 \times 1 = 10$
24. $0 \times 5 = 0$ chores
25. $6 \times 1 = 6$ plates

Lesson Practice 2C

1. $6 \times 0 = 0$
2. $0 \times 2 = 0$
3. $5 \times 1 = 5$
4. $1 \times 8 = 8$
5. $1 \times 7 = 7$
6. $6 \times 1 = 6$
7. $1 \times 2 = 2$
8. $9 \times 1 = 9$
9. $1 \times 5 = 5$
10. $4 \times 1 = 4$
11. $1 \times 0 = 0$
12. $6 \times 0 = 0$
13. $4 \times 0 = 0$
14. $0 \times 3 = 0$
15. $0 \times 8 = 0$
16. $0 \times 2 = 0$
17. $3 \times 1 = 3$
18. $7 \times 0 = 0$
19. $1 \times 4 = 4$
20. $1 \times 8 = 8$
21. $9 \times 1 = 9$
    $1 \times 9 = 9$
22. $0 + 0 + 0 = 0$
23. $9 = 9$
24. $1 \times 8 = 8$ noses
25. $7 \times 0 = 0$ grapes

Systematic Review 2D

1. $1 \times 9 = 9$
2. $3 \times 1 = 3$

3. $7 \times 0 = 0$
4. $9 \times 0 = 0$
5. $1 \times 0 = 0$
6. $1 \times 6 = 6$
7. $1 \times 5 = 5$
8. $1 \times 1 = 1$
9. $0 \times 2 = 0$
10. $0 \times 3 = 0$
11. $0 \times 1 = 0$
12. $7 \times 1 = 7$
13. $4 \times 1 = 4$
14. $2 \times 1 = 2$
15. $0 \times 8 = 0$
16. $6 \times 0 = 0$
17. $5 + 7 = 12$
18. $3 + 4 = 7$
19. $7 - 5 = 2$
20. $5 + 4 = 9$
21. $4 - 0 = 4$
22. $6 + 6 = 12$
23. $6 + 3 = 9$
24. $15 - 9 = 6$
25. $0 + 0 + 0 + 0 + 0 + 0 + 0 = 0$
26. $1 \times 8 = 8$ bunnies

Systematic Review 2E

1. $0 \times 6 = 0$
2. $8 \times 0 = 0$
3. $2 \times 1 = 2$
4. $1 \times 4 = 4$
5. $0 \times 2 = 0$
6. $0 \times 3 = 0$
7. $1 \times 0 = 0$
8. $7 \times 1 = 7$
9. $8 \times 1 = 8$
10. $0 \times 5 = 0$
11. $4 \times 1 = 4$
12. $5 \times 0 = 0$
13. $0 \times 1 = 0$
14. $6 \times 1 = 6$
15. $5 \times 1 = 5$
16. $1 \times 1 = 1$

17. $5+3=8$
18. $8+6=14$
19. $15-7=8$
20. $7+3=10$
21. $12-7=5$
22. $7+6=13$
23. $6+5=11$
24. $11-3=8$
25. $3=3$
26. $8\times1=8$ coats

Systematic Review 2F

1. $1\times10=10$
2. $4\times1=4$
3. $6\times0=0$
4. $1\times7=7$
5. $0\times8=0$
6. $9\times1=9$
7. $5\times1=5$
8. $7\times0=0$
9. $6\times1=6$
10. $1\times1=1$
11. $0\times3=0$
12. $10\times0=0$
13. $16-8=8$
14. $4+5=9$
15. $7-4=3$
16. $8+2=10$
17. $6+4=10$
18. $12-5=7$
19. $5+6=11$
20. $3+6=9$
21. $13-5=8$
22. $8-0=8$
23. $3+9=12$
24. $17-8=9$
25. $0+0=0$
26. $0\times4=0$ cans

Lesson Practice 3A

1. 2,4,6,8,10,12,
   14,16,18,20
2. 10,20,30,40,50,
   60,70,80,90,100
3. 5,10,15,20,25,
   30,35,40,45,50
4. 5,10,15,20,25,
   30,35,40,45,50
5. 2,4,6,8,10,12,
   14,16,18,20
6. 5,10,15,20,25,
   30,35,40,45,50
7. 10,20,30,40,50,
   60,70,80,90,100

Lesson Practice 3B

1. 2,4,6,8,10,12,
   14,16,18,20
2. 10,20,30,40,50,
   60,70,80,90,100
3. 5,10,15,20,25,
   30,35,40,45,50
4. 5,10,15,20,25,
   30,35,40,45,50
5. 2,4,6,8,10,12,
   14,16,18,20
6. 5,10,15,20,25,
   30,35,40,45,50
7. 10,20,30,40,50,
   60,70,80,90,100

Lesson Practice 3C

1. 2,4,6,8,10,12,
   14,16,18,20
2. 10,20,30,40,50,
   60,70,80,90,100
3. 5,10,15,20,25,
   30,35,40,45,50

4. 5,10,15,20,25,
   30,35,40,45,50
5. 2,4,6,8,10,12,
   14,16,18,20
6. 5,10,15,20,25,
   30,35,40,45,50
7. 10,20,30,40,50,
   60,70,80,90,100

8. $0 \times 3 = 0$
9. $1 \times 1 = 1$
10. $0 \times 8 = 0$
11. $4 \times 1 = 4$
12. $8 - 4 = 4$
13. $10 + 3 = 13$
14. $13 - 6 = 7$
15. $5 + 7 = 12$
16. 0
17. 10, 20, 30, 40, 50; 5 bags
18. $7 \times 1 = 7$ times
19. $11 - 6 = 5$ snacks
20. $\$7 + \$3 = \$10$
    $\$10 - \$1 = \$9$

Systematic Review 3D

1. 5, 10, 15, 20, 25, 30, 35, 40, 45, 50
2. 10, 20, 30, 40, 50, 60, 70, 80, 90, 100
3. 2, 4, 6, 8, 10, 12, 14, 16, 18, 20
4. $1 \times 7 = 7$
5. $6 \times 1 = 6$
6. $10 \times 0 = 0$
7. $0 \times 2 = 0$
8. $3 \times 1 = 3$
9. $8 \times 1 = 8$
10. $0 \times 4 = 0$
11. $0 \times 0 = 0$
12. $7 + 8 = 15$
13. $14 \quad 5 - 9$
14. $12 - 4 = 8$
15. $8 + 9 = 17$
16. $0+0+0+0+0+0+0+0+0 = 0$
17. 5, 10, 15, 20, 25, 30, 35; 7 flowers
18. $7 \times 1 = 7$ cookies
19. $6 + 5 = 11$ chapters
20. $7 + 9 = 16$
    $16 \times 0 = 0$ rocket ships

Systematic Review 3E

1. 5, 10, 15, 20, 25, 30, 35, 40, 45, 50
2. 2, 4, 6, 8, 10, 12, 14, 16, 18, 20
3. 10, 20, 30, 40, 50, 60, 70, 80, 90, 100
4. $5 \times 1 = 5$
5. $6 \times 0 = 0$
6. $1 \times 10 = 10$
7. $9 \times 1 = 9$

Systematic Review 3F

1. 10, 20, 30, 40, 50, 60, 70, 80, 90, 100
2. 5, 10, 15, 20, 25, 30, 35, 40, 45, 50
3. 2, 4, 6, 8, 10, 12, 14, 16, 18, 20
4. $1 \times 0 = 0$
5. $2 \times 1 = 2$
6. $5 \times 0 = 0$
7. $8 \times 1 = 8$
8. $6 \times 1 = 6$
9. $10 \times 0 = 0$
10. $7 \times 1 = 7$
11. $9 \times 0 = 0$
12. $6 + 7 = 13$
13. $18 - 9 = 9$
14. $13 - 4 = 9$
15. $8 + 5 = 13$
16. $5 = 5$
17. 2, 4, 6, 8, 10, 12, 14, 16, 18 mittens
18. $10 \times 1 = 10$ kites
19. $9 - 2 = 7$
    $7 - 2 = 5$ apples
20. $0 \times 5 = 0$
    $1 \times 6 = 6$
    $0 + 6 = 6$ stories

Lesson Practice 4A

1. $1 \times 2 = 2$
2. $2 \times 3 = 6$
3. $2 \times 5 = 10$
4. $2 \times 7 = 14$
5. $9 \times 2 = 18$
6. $2 \times 2 = 4$
7. $2 \times 10 = 20$
8. $8 \times 2 = 16$
9. $4 \times 2 = 8$
10. $7 \times 2 = 14$
11. $2 \times 6 = 12$
12. $2 \times 0 = 0$
13. $4 \times 2 = 8$
    $2 \times 4 = 8$

14,15. $\dfrac{0}{2 \times 0}$ $\dfrac{2}{2 \times 1}$ $\dfrac{4}{2 \times 2}$ $\dfrac{6}{2 \times 3}$ $\dfrac{8}{2 \times 4}$ $\dfrac{10}{2 \times 5}$
$\dfrac{12}{2 \times 6}$ $\dfrac{14}{2 \times 7}$ $\dfrac{16}{2 \times 8}$ $\dfrac{18}{2 \times 9}$ $\dfrac{20}{2 \times 10}$

16. $6 \times 2 = 12$ pints
17. 8×2
18. $2+2+2+2+2+2+2+2+2+2 = 20$
19. $10 \times 2 = 20$ pages
20. $7 \times 2 = 14$ pints

Lesson Practice 4B

1. $2 \times 2 = 4$
2. $2 \times 6 = 12$
3. $2 \times 8 = 16$
4. $9 \times 2 = 18$
5. $2 \times 7 = 14$
6. $4 \times 2 = 8$
7. $1 \times 2 = 2$
8. $3 \times 2 = 6$
9. $0 \times 2 = 0$
10. $5 \times 2 = 10$
11. $2 \times 8 = 16$
12. $10 \times 2 = 20$
13. $9 \times 2 = 18$
    $2 \times 9 = 18$

14,15. $\dfrac{0}{2 \cdot 0}$ $\dfrac{2}{2 \cdot 1}$ $\dfrac{4}{2 \cdot 2}$ $\dfrac{6}{2 \cdot 3}$ $\dfrac{8}{2 \cdot 4}$ $\dfrac{10}{2 \cdot 5}$
$\dfrac{12}{2 \cdot 6}$ $\dfrac{14}{2 \cdot 7}$ $\dfrac{16}{2 \cdot 8}$ $\dfrac{18}{2 \cdot 9}$ $\dfrac{20}{2 \cdot 10}$

16. $8 \times 2 = 16$ pints
17. 6×2
18. $2+2+2+2+2+2+2 = 14$
19. $3 \times 2 = 6$ children
20. $8 \times 2 = 16$ pints

Lesson Practice 4C

1. $9 \times 2 = 18$
2. $2 \times 4 = 8$
3. $2 \times 7 = 14$
4. $8 \times 2 = 16$
5. $5 \times 2 = 10$
6. $1 \times 2 = 2$
7. $2 \times 6 = 12$
8. $10 \times 2 = 20$
9. $2 \times 2 = 4$
10. $3 \times 2 = 6$
11. $2 \times 9 = 18$
12. $0 \times 2 = 0$
13. $2 \times 2 = 4$

14,15. $\dfrac{0}{(2)(0)},$ $\dfrac{2}{(2)(1)},$ $\dfrac{4}{(2)(2)},$ $\dfrac{6}{(2)(3)},$
$\dfrac{8}{(2)(4)},$ $\dfrac{10}{(2)(5)},$ $\dfrac{12}{(2)(6)},$ $\dfrac{14}{(2)(7)},$
$\dfrac{16}{(2)(8)},$ $\dfrac{18}{(2)(9)},$ $\dfrac{20}{(2)(10)}$

16. $10 \times 2 = 20$ pints
17. 7×2
18. $2+2 = 4$
19. $9 \times 2 = 18$ pints
20. $6 \times 2 = 12$ rabbits

Systematic Review 4D

1. $5 \times 2 = 10$
2. $2 \times 6 = 12$
3. $9 \times 2 = 18$
4. $2 \times 8 = 16$

5. $1 \times 3 = 3$
6. $2 \times 7 = 14$
7. $0 \times 6 = 0$
8. $4 \times 2 = 8$
9. $3 \times 2 = 6$
   $2 \times 3 = 6$
10. $10 \times 2 = 20$
    $2 \times 10 = 20$
11. $2 \times 1 = 2$
    $1 \times 2 = 2$
12. $0 \times 2 = 0$
    $2 \times 0 = 0$
13. 2, 4, 6, 8, 10,
    12, 14, 16, 18, 20
14. $12 - 4 = 8$
15. $9 + 8 = 17$
16. $15 - 7 = 8$
17. $5 + 4 = 9$
18. $500 + 40 + 2$
19. $100 + 60 + 3$
20. $2 + 2 + 2 + 2 + 2 + 2 + 2 = 14$
21. $2 \times 2 = 4$ pints
22. $5 + 3 = 8$ children
    $2 \times 8 = 16$ wheels

Systematic Review 4E

1. $0 \times 2 = 0$
2. $5 \times 2 = 10$
3. $2 \times 2 = 4$
4. $2 \times 4 = 8$
5. $2 \times 3 = 6$
6. $9 \times 1 = 9$
7. $2 \times 6 = 12$
8. $10 \times 2 = 20$
9. $7 \times 2 = 14$
   $2 \times 7 = 14$
10. $8 \times 2 = 16$
    $2 \times 8 = 16$
11. $5 \times 1 = 5$
    $1 \times 5 = 5$
12. $2 \times 6 = 12$
    $6 \times 2 = 12$

13. 10, 20, 30, 40, 50,
    60, 70, 80, 90, 100
14. $16 - 8 = 8$
15. $5 + 3 = 8$
16. $18 - 9 = 9$
17. $7 + 5 = 12$
18. $300 + 50 + 1$
19. $200 + 40 + 9$
20. $7 + 7 = 14$
21. $2 \times 8 = 16$ mittens
22. $5 \times 2 = 10$
    $10 - 3 = 7$ eggs

Systematic Review 4F

1. $3 \times 2 = 6$
2. $2 \times 10 = 20$
3. $8 \times 2 = 16$
4. $1 \times 7 = 7$
5. $2 \times 3 = 6$
6. $2 \times 6 = 12$
7. $4 \times 2 = 8$
8. $0 \times 9 = 0$
9. $5 \times 2 = 10$
   $2 \times 5 - 10$
10. $7 \times 2 = 14$
    $2 \times 7 = 14$
11. $9 \times 2 = 18$
    $2 \times 9 = 18$
12. $4 \times 1 = 4$
    $1 \times 4 = 4$
13. 5, 10, 15, 20, 25, 30, 35, 40, 45, 50
14. $16 - 9 = 7$
15. $7 + 7 = 14$
16. $9 - 4 = 5$
17. $5 + 6 = 11$
18. $100 + 30 + 1$
19. $400 + 70 + 5$
20. $2 \times 10 = 20$ peanuts
21. $3 \times 2 = 6$ pints
22. $2 \times 2 = 4$ hats
    $3 \times 2 = 6$ hats
    $4 + 6 = 10$ hats

Lesson Practice 5A

1. $10 \times 0 = 0$
2. $5 \times 10 = 50$
3. $10 \times 2 = 20$
4. $6 \times 10 = 60$
5. $10 \times 10 = 100$
6. $10 \times 3 = 30$
7. $10 \times 9 = 90$
8. $10 \times 7 = 70$
9. $10 \times 2 = 20$
10. $10 \times 5 = 50$
11. $10 \times 1 = 10$
12. $10 \times 3 = 30$
13. $10 \times 7 = 70$
    $7 \times 10 = 70$
14. $4 \times 10 = 40$
    $10 \times 4 = 40$
15. $10 \times 6 = 60$
    $6 \times 10 = 60$
16. $10 \times 3 = 30$
    $3 \times 10 = 30$

17.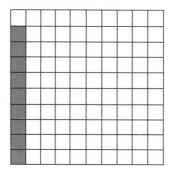

18. $10 + 10 + 10 + 10 = 40¢$ or 40 pennies
19. $10 + 10 + 10 + 10 + 10 +$
    $10 + 10 + 10 + 10 = 90$
20. $10 \times 6 = 60$ cars

Lesson Practice 5B

1. $10 \times 8 = 80$
2. $1 \times 10 = 10$
3. $10 \times 9 = 90$
4. $0 \times 10 = 0$
5. $10 \times 5 = 50$
6. $10 \times 4 = 40$
7. $10 \times 6 = 60$
8. $10 \times 10 = 100$
9. $10 \times 8 = 80$
10. $10 \times 7 = 70$
11. $10 \times 2 = 20$
12. $10 \times 1 = 10$
13. $10 \times 5 = 50$
    $5 \times 10 = 50$
14. $8 \times 10 = 80$
    $10 \times 8 = 80$
15. $10 \times 0 = 0$
    $0 \times 10 = 0$
16. $10 \times 9 = 90$
    $9 \times 10 = 90$

17.

0	10	20	30
(10)(0)	(10)(1)	(10)(2)	(10)(3)

40	50	60	70
(10)(4)	(10)(5)	(10)(6)	(10)(7)

80	90	100
(10)(8)	(10)(9)	(10)(10)

18. $10 + 10 + 10 + 10 + 10 + 10 + 10 = 70¢$
19. $10 \times 6 = 60$
20. $10 \times 5 = 50$ problems

Lesson Practice 5C

1. $3 \times 10 = 30$
2. $8 \times 10 = 80$
3. $10 \times 1 = 10$
4. $2 \times 10 = 20$
5. $10 \times 9 = 90$
6. $7 \times 10 = 70$
7. $10 \times 5 = 50$
8. $6 \times 10 = 60$
9. $10 \times 0 = 0$
10. $10 \times 4 = 40$
11. $10 \times 10 = 100$
12. $10 \times 3 = 30$
13. $10 \times 1 = 10$
    $1 \times 10 = 10$
14. $10 \times 4 = 40$
    $4 \times 10 = 40$

15. $10 \times 2 = 20$
    $2 \times 10 = 20$
16. $7 \times 10 = 70$
    $10 \times 7 = 70$
17. see 5A #17
18. $10+10+10+10+10 = 50¢$
19. $10 \times 3 = 30$
20. $\$10 \times 2 = \$20$

Systematic Review 5D

1. $10 \times 5 = 50$
2. $7 \times 10 = 70$
3. $10 \times 2 = 20$
4. $10 \times 10 = 100$
5. $2 \times 5 = 10$
6. $10 \times 5 = 50$
7. $6 \times 2 = 12$
8. $7 \times 2 = 14$
9. $1 \times 3 = 3$
10. $9 \times 2 = 18$
11. $10 \times 8 = 80$
12. $10 \times 4 = 40$
13. $9 \times 2 = 18$
    $2 \times 9 = 18$
14. $4 \times 2 = 8$
    $2 \times 4 = 8$
15. $10 \times 3 = 30$
    $3 \times 10 = 30$
16. $5 \times 2 = 10$
    $2 \times 5 = 10$
17. done
18. $\begin{array}{r} 43 \\ +43 \\ \hline 86 \end{array}$
19. $\begin{array}{r} 28 \\ -16 \\ \hline 12 \end{array}$
20. $\begin{array}{r} 89 \\ -51 \\ \hline 38 \end{array}$
21. $7 \times 10 = 70$ hours
22. $70+20 = 90$ hours

Systematic Review 5E

1. $10 \times 8 = 80$
2. $6 \times 10 = 60$
3. $10 \times 9 = 90$
4. $10 \times 0 = 0$
5. $5 \times 1 = 5$
6. $6 \times 2 = 12$
7. $8 \times 1 = 8$
8. $10 \times 5 = 50$
9. $2 \times 2 = 4$
10. $2 \times 5 = 10$
11. $9 \times 1 = 9$
    $1 \times 9 = 9$
12. $3 \times 10 = 30$
    $10 \times 3 = 30$
13. $300+80+9$
14. $70+2$
15. $\begin{array}{r} 46 \\ +22 \\ \hline 68 \end{array}$
16. $\begin{array}{r} 51 \\ +12 \\ \hline 63 \end{array}$
17. $\begin{array}{r} 37 \\ -23 \\ \hline 14 \end{array}$
18. $\begin{array}{r} 94 \\ -43 \\ \hline 51 \end{array}$
19. $10+10+10+10+$
    $10+10+10+10 = 80¢$
20. $4 \times 10 = 40$ fingers
21. $6+4 = 10$
    $10 \times 10 = 100$ pieces
22. $9 \times 2 = 18$ pints

Systematic Review 5F

1. $4 \times 1 = 4$
2. $2 \times 10 = 20$
3. $10 \times 3 = 30$
4. $10 \times 9 = 90$

5. $6 \times 2 = 12$

6. $2 \times 8 = 16$

7. $10 \times 7 = 70$

8. $10 \times 1 = 10$

9. $3 \times 2 = 6$

10. $4 \times 2 = 8$

11. $1 \times 6 = 6$

12. $9 \times 0 = 0$

13. $100 + 60 + 4$

14. $50 + 8$

15.
$$\begin{array}{r} 52 \\ -20 \\ \hline 32 \end{array}$$

16.
$$\begin{array}{r} 64 \\ +13 \\ \hline 77 \end{array}$$

17.
$$\begin{array}{r} 35 \\ +34 \\ \hline 69 \end{array}$$

18.
$$\begin{array}{r} 14 \\ -12 \\ \hline 2 \end{array}$$

19. $5+5+5+5+5+5+5+5+5+5 = 50$

20. $9 \times 10 = 90¢$

21. Wayne: $\$5 \times 10 = \$50$
Together: $\$50 + \$5 = \$55$

22. $2 \times 8 = 16$ pints

Lesson Practice 6A

1. $5 \times 4 = 20$

2. $5 \times 9 = 45$

3. $5 \times 8 = 40$

4. $5 \times 10 = 50$

5. $2 \times 5 = 10$

6. $5 \times 5 = 25$

7. $5 \times 1 = 5$

8. $5 \times 3 = 15$

9. $7 \times 5 = 35$

10. $0 \times 5 = 0$

11. $6 \times 5 = 30$

12. $5 \times 5 = 25$

13. $5 \times 10 = 50$
$10 \times 5 = 50$

14. $5 \times 7 = 35$
$7 \times 5 = 35$

15. $5 \times 3 = 15$
$3 \times 5 = 15$

16. $5 \times 6 = 30$
$6 \times 5 = 30$

17.
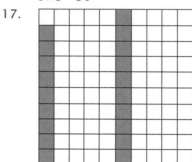

18. $5 + 5 = 10¢$

19. $5 + 5 + 5 + 5 = 20$

20. $5 \times 8 = 40$ fingers

Lesson Practice 6B

1. $5 \times 8 = 40$

2. $5 \times 4 = 20$

3. $5 \times 6 = 30$

4. $5 \times 1 = 5$

5. $2 \times 5 = 10$

6. $9 \times 5 = 45$

7. $5 \times 3 = 15$

8. $7 \times 5 = 35$

9. $5 \times 5 = 25$

10. $5 \times 0 = 0$

11. $10 \times 5 = 50$

12. $5 \times 4 = 20$

13. $5 \times 2 = 10$
$2 \times 5 = 10$

14. $5 \times 8 = 40$
$8 \times 5 = 40$

15. $9 \times 5 = 45$
$5 \times 9 = 45$

16. $5 \times 1 = 5$

    $1 \times 5 = 5$

17. $6 \times 5 = 30$

    $5 \times 6 = 30$

18.

$\underline{0}$	$\underline{5}$	$\underline{10}$	$\underline{15}$	$\underline{20}$	$\underline{25}$
5×0	5×1	5×2	5×3	5×4	5×5

$\underline{30}$	$\underline{35}$	$\underline{40}$	$\underline{45}$	$\underline{50}$
5×6	5×7	5×8	5×9	5×10

19. $5+5+5+5+5+5+5+5 = 40$

20. $5 \times 5 \text{¢} = 25 \text{¢}$

Lesson Practice 6C

1. $5 \times 2 = 10$
2. $6 \times 5 = 30$
3. $5 \times 10 = 50$
4. $0 \times 5 = 0$
5. $5 \times 1 = 5$
6. $7 \times 5 = 35$
7. $5 \times 5 = 25$
8. $4 \times 5 = 20$
9. $5 \times 3 = 15$
10. $8 \times 5 = 40$
11. $5 \times 9 = 45$
12. $5 \times 6 = 30$
13. $5 \times 4 = 20$

    $4 \times 5 = 20$

14. $5 \times 10 = 50$

    $10 \times 5 = 50$

15. $5 \times 7 = 35$

    $7 \times 5 = 35$

16. $5 \times 0 = 0$

    $0 \times 5 = 0$

17. see 6A #17
18. $5 \times 8 = 40\text{¢}$
19. $5 \times 10 = 50$
20. $5 \times 7 = 35$ pages

Systematic Review 6D

1. $5 \times 6 = 30$
2. $10 \times 7 = 70$

3. $2 \times 6 = 12$
4. $5 \times 8 = 40$
5. $9 \times 5 = 45$
6. $10 \times 6 = 60$
7. $8 \times 2 = 16$
8. $7 \times 5 = 35$
9. $0 \times 3 = 0$
10. $9 \times 1 = 9$
11. $5 \times 4 = 20$
12. $10 \times 3 = 30$
13. $3 \times 5 = 15$

    $5 \times 3 = 15$

14. $7 \times 2 = 14$

    $2 \times 7 = 14$

15. $10 \times 5 = 50$

    $5 \times 10 = 50$

16. $1 \times 2 = 2$

    $2 \times 1 = 2$

17.
$$\begin{array}{r} {}^1 25 \\ +36 \\ \hline 61 \end{array}$$

18.
$$\begin{array}{r} 1\,{}^1 78 \\ +34 \\ \hline 112 \end{array}$$

19.
$$\begin{array}{r} 1\,{}^1 49 \\ +51 \\ \hline 100 \end{array}$$

20.
$$\begin{array}{r} {}^1 65 \\ +15 \\ \hline 80 \end{array}$$

21. $5 \times \underline{7} = 35\text{¢}$
22. $25 + 58 = 83$ minutes

Systematic Review 6E

1. $5 \times 5 = 25$
2. $1 \times 5 = 5$
3. $2 \times 9 = 18$
4. $10 \times 10 = 100$
5. $10 \times 8 = 80$
6. $5 \times 2 = 10$
7. $6 \times 5 = 30$
8. $9 \times 5 = 45$

9. $7 \times 1 = 7$

10. $2 \times 3 = 6$

11. $8 \times 2 = 16$

12. $9 \times 0 = 0$

13. $4 \times 5 = 20$

$5 \times 4 = 20$

14. $10 \times 2 = 20$

$2 \times 10 = 20$

15. $5 \times 7 = 35$

$7 \times 5 = 35$

16. $5 \times 3 = 15$

$3 \times 5 = 15$

17.
$$\begin{array}{r} {}^1 27 \\ +34 \\ \hline 61 \end{array}$$

18.
$$\begin{array}{r} {}^1 19 \\ +13 \\ \hline 32 \end{array}$$

19.
$$\begin{array}{r} {}^1 61 \\ +29 \\ \hline 90 \end{array}$$

20.
$$\begin{array}{r} {}^1 47 \\ +37 \\ \hline 84 \end{array}$$

21. $5 + 5 + 5 = 15$

22. $9 \times 5 = 45$ miles

23. $18 + 19 = 37$ experiments

24. $15 - 5 = 10$

$10 + 15 = 25$ daisies

12. $5 \times 7 = 35$

13.
$$\begin{array}{r} 61 \\ -30 \\ \hline 31 \end{array}$$

14.
$$\begin{array}{r} {}^1 28 \\ +23 \\ \hline 51 \end{array}$$

15.
$$\begin{array}{r} {}^1 49 \\ +14 \\ \hline 63 \end{array}$$

16.
$$\begin{array}{r} 35 \\ +64 \\ \hline 99 \end{array}$$

17.
$$\begin{array}{r} {}^1 57 \\ +27 \\ \hline 84 \end{array}$$

18.
$$\begin{array}{r} 24 \\ -13 \\ \hline 11 \end{array}$$

19.
$$\begin{array}{r} 88 \\ -24 \\ \hline 64 \end{array}$$

20.
$$\begin{array}{r} {}^1 83 \\ +9 \\ \hline 92 \end{array}$$

21. $9 + 9 + 9 + 9 + 9 = 45$

22. $8 \times 5 = 40¢$

23. $8 \times 2 = 16$ pints

24. $43 + 29 = 72$ lights

$72 - 10 = 62$ bulbs

Systematic Review 6F

1. $5 \times 0 = 0$

2. $5 \times 10 = 50$

3. $8 \times 5 = 40$

4. $9 \times 10 = 90$

5. $10 \times 4 = 40$

6. $2 \times 6 = 12$

7. $5 \times 2 = 10$

8. $5 \times 3 = 15$

9. $6 \times 0 = 0$

10. $8 \times 1 = 8$

11. $2 \times 4 = 8$

Lesson Practice 7A

1. done

2. $5 \times 5 = 25$

3. $2 \times 2 = 4$ sq in

4. $2 \times 3 = 6$ sq in

5. $5 \times 2 = 10$ sq ft

6. $1 \times 1 = 1$ sq mi

7. $5 \times 8 = 40$ sq ft

8. $10 \times 9 = 90$ tiles

Lesson Practice 7B

1. done
2. $10 \times 5 = 50$ sq in
3. $2 \times 4 = 8$ sq in
4. $5 \times 7 = 35$ sq ft
5. $10 \times 10 = 100$ sq ft
6. $2 \times 9 = 18$ sq in
7. $5 \times 5 = 25$ sq mi
8. $5 \times 6 = 30$ sq in

Lesson Practice 7C

1. $6 \times 1 = 6$ sq ft
2. $2 \times 2 = 4$ sq in
3. $10 \times 2 = 20$ sq mi
4. $10 \times 4 = 40$ sq ft
5. $5 \times 9 = 45$ sq in
6. $5 \times 4 = 20$ sq mi
7. 10 ft $\times 10$ ft $= 100$ sq ft; yes
8. $8 \times 2 = 16$ blocks

Systematic Review 7D

1. $10 \times 8 = 80$ sq in
2. $3 \times 5 = 15$ sq ft
3. $1 \times 1 = 1$ sq mi
4. $5 \times 9 = 45$
5. $10 \times 6 = 60$
6. $5 \times 7 = 35$
7. $0 \times 6 = 0$
8. $1 \times 8 = 8$
9. $7 \times 2 = 14$
10. $2 \times 5 = 10$
11. $5 \times 6 = 30$
12.
$$\begin{array}{r} {}^1 35 \\ +56 \\ \hline 91 \end{array}$$
13.
$$\begin{array}{r} 48 \\ -24 \\ \hline 24 \end{array}$$

14.
$$\begin{array}{r} {}^1 78 \\ +\ 9 \\ \hline 87 \end{array}$$
15.
$$\begin{array}{r} 84 \\ -13 \\ \hline 71 \end{array}$$
16. $2 \times 9 = 18$ sq mi
17. $6 \times 2 = 12$ pints
18. 3 ft $\times 2$ ft $= 6$ sq ft
    $6 \times 2 = 12$ cats

Systematic Review 7E

1. $10 \times 7 = 70$ sq ft
2. $5 \times 5 = 25$ sq mi
3. $2 \times 7 = 14$ sq in
4. $1 \times 9 = 9$
5. $10 \times 4 = 40$
6. $5 \times 8 = 40$
7. $2 \times 6 = 12$
8. $10 \times 1 = 10$
9. $5 \times 3 = 15$
10. $10 \times 7 = 70$
11. $9 \times 2 = 18$
12.
$$\begin{array}{r} 79 \\ -21 \\ \hline 58 \end{array}$$
13.
$$\begin{array}{r} {}^1 32 \\ +59 \\ \hline 91 \end{array}$$
14.
$$\begin{array}{r} 63 \\ -30 \\ \hline 33 \end{array}$$
15.
$$\begin{array}{r} {}^1 45 \\ +45 \\ \hline 90 \end{array}$$
16. $6 \times 5¢ = 30¢$
17. $3 \times 5 = 15$ sq ft
18. $16 + 26 = 42$ acorns
    $42 - 12 = 30$ acorns

Systematic Review 7F

1. $4 \times 1 = 4$ sq in
2. $10 \times 10 = 100$ sq mi
3. $5 \times 9 = 45$ sq units
4. $10 \times 3 = 30$
5. $5 \times 2 = 10$
6. $4 \times 5 = 20$
7. $10 \times 2 = 20$
8. $1 \times 1 = 1$
9. $8 \times 2 = 16$
10. $2 \times 4 = 8$
11. $5 \times 7 = 35$
12. $\begin{array}{r} {}^{1}{}^{1}67 \\ +53 \\ \hline 120 \end{array}$
13. $\begin{array}{r} {}^{1}19 \\ +12 \\ \hline 31 \end{array}$
14. $\begin{array}{r} {}^{1}93 \\ +8 \\ \hline 101 \end{array}$
15. $\begin{array}{r} 61 \\ -40 \\ \hline 21 \end{array}$
16. $50 + 50 = 100$
    $10 \times 10 = 100$ sq ft; yes
17. 5, 10, 15, 20, 25, 30, 35, 40, 45; 9 rows
18. $13 + 12 = 25$
    $25 - 5 = 20$ pt
    2, 4, 6, 8, 10, 12, 14, 16, 18, 20
    10 qt of jelly left

Lesson Practice 8A

1. done
2. $5 \times \underline{5} = 25$
3. $10 \times \underline{8} = 80$
4. $1 \times \underline{7} = 7$
5. $4 \times \underline{5} = 20$
6. $10 \times \underline{5} = 50$
7. $2 \times \underline{9} = 18$
8. $5 \times \underline{3} = 15$

9. $3 \times \underline{1} = 3$
10. $2 \times \underline{2} = 4$
11. $1 \times \underline{10} = 10$
12. $8 \times \underline{2} = 16$
13. $7 \times \underline{2} = 14$
14. $5 \times \underline{6} = 30$
15. $10 \times \underline{10} = 100$
16. $5 \times \underline{9} = 45$
17. $2 \times \underline{6} = 12$; 6 people
18. $2 \times \underline{9} = 18$; 9 gumdrops
19. $10 \times \underline{4} = 40$; 4 vases
20. $5 \times \underline{10} = 50$; 10 weeks

Lesson Practice 8B

1. $2 \times \underline{10} = 20$
2. $5 \times \underline{7} = 35$
3. $10 \times \underline{9} = 90$
4. $6 \times \underline{0} = 0$
5. $4 \times \underline{2} = 8$
6. $7 \times \underline{10} = 70$
7. $5 \times \underline{6} = 30$
8. $8 \times \underline{1} = 8$
9. $2 \times \underline{6} = 12$
10. $10 \times \underline{6} = 60$
11. $8 \times \underline{5} = 40$
12. $10 \times \underline{2} = 20$
13. $3 \times \underline{2} = 6$
14. $1 \times \underline{4} = 4$
15. $2 \times \underline{10} = 20$
16. $8 \times \underline{0} = 0$
17. $2 \times \underline{7} = 14$; 7 days
18. $2 \times \underline{8} = 16$; 8 people
19. $5 \times \underline{7} = 35$; 7 cages
20. $10 \times \underline{\$10} = \$100$; $10 each

Lesson Practice 8C

1. $2 \times 6 = 12$
2. $5 \times 2 = 10$
3. $10 \times 4 = 40$
4. $1 \times 6 = 6$
5. $5 \times 5 = 25$
6. $8 \times 10 = 80$
7. $9 \times 1 = 9$
8. $5 \times 9 = 45$
9. $2 \times 8 = 16$
10. $3 \times 5 = 15$
11. $10 \times 10 = 100$
12. $9 \times 2 = 18$
13. $3 \times 10 = 30$
14. $5 \times 8 = 40$
15. $5 \times 0 = 0$
16. $2 \times 1 = 2$
17. $2 \times 5 = 10$; 5 children
18. $5 \times 4 = 20$; 4 cars
19. $2 \times 4 = 8$; 4 qt
20. $\$10 \times 9 = \$90$; 9 bills

Systematic Review 8D

1. $2 \times 10 = 20$
2. $10 \times 6 = 60$
3. $5 \times 7 = 35$
4. $5 \times 1 = 5$
5. $7 \times 2 = 14$
6. $6 \times 5 = 30$
7. $5 \times 1 = 5$ sq in
8. $5 \times 5 = 25$ sq ft
9. $5 \times 4 = 20$ sq mi

10.
$$\begin{array}{r} {}^{1}6\,1 \\ +\,2\,9 \\ \hline 9\,0 \end{array}$$

11.
$$\begin{array}{r} {}^{1}7\,2 \\ +\,3\,8 \\ \hline 1\,1\,0 \end{array}$$

12.
$$\begin{array}{r} 4\,4 \\ +\,5\,5 \\ \hline 9\,9 \end{array}$$

13.
$$\begin{array}{r} 8\,6 \\ +\,7\,3 \\ \hline 1\,5\,9 \end{array}$$

14.
$$\begin{array}{r} {}^{2}\!\not{3}\,{}^{1}4 \\ -\,1\,6 \\ \hline 1\,8 \end{array}$$

15.
$$\begin{array}{r} {}^{6}\!\not{7}\,{}^{1}4 \\ -\,3\,8 \\ \hline 3\,6 \end{array}$$

16.
$$\begin{array}{r} {}^{6}\!\not{7}\,{}^{1}1 \\ -\,5\,9 \\ \hline 1\,2 \end{array}$$

17.
$$\begin{array}{r} 6\,7 \\ -\,2\,5 \\ \hline 4\,2 \end{array}$$

18. $65 - 37 = 28$ cookies

Systematic Review 8E

1. $5 \times 8 = 40$
2. $10 \times 7 = 70$
3. $2 \times 9 = 18$
4. $1 \times 1 = 1$
5. $3 \times 2 = 6$
6. $9 \times 5 = 45$
7. $10 \times 4 = 40$ sq in
8. $2 \times 2 = 4$ sq ft
9. $10 \times 8 = 80$ sq ft

10.
$$\begin{array}{r} {}^{1}3\,5 \\ +\,6\,5 \\ \hline 1\,0\,0 \end{array}$$

11.
$$\begin{array}{r} 5\,3 \\ +\,6\,1 \\ \hline 1\,1\,4 \end{array}$$

12.
$$\begin{array}{r} {}^{1}9\,9 \\ +\,2\,2 \\ \hline 1\,2\,1 \end{array}$$

13.
$$\begin{array}{r} {}^{1}26 \\ +48 \\ \hline 74 \end{array}$$

14.
$$\begin{array}{r} 17 \\ -\ 9 \\ \hline 8 \end{array}$$

15.
$$\begin{array}{r} {}^{3}4\ {}^{1}5 \\ -\ 2\ 6 \\ \hline 1\ 9 \end{array}$$

16.
$$\begin{array}{r} {}^{8}9\ {}^{1}3 \\ -\ 3\ 4 \\ \hline 5\ 9 \end{array}$$

17.
$$\begin{array}{r} 52 \\ -42 \\ \hline 10 \end{array}$$

18. $5 \times 2 = 10$ pints

19. $2 \times 6 = 12$; 6 peanuts

20. $10 \times 10 = 100$ sq mi

13.
$$\begin{array}{r} {}^{1}76 \\ +75 \\ \hline 151 \end{array}$$

14.
$$\begin{array}{r} {}^{1}2\ {}^{1}4 \\ -\ 1\ 5 \\ \hline 9 \end{array}$$

15.
$$\begin{array}{r} {}^{4}5\ {}^{1}3 \\ -\ 3\ 5 \\ \hline 1\ 8 \end{array}$$

16.
$$\begin{array}{r} {}^{5}6\ {}^{1}5 \\ -\ 1\ 9 \\ \hline 4\ 6 \end{array}$$

17.
$$\begin{array}{r} {}^{2}3\ {}^{1}7 \\ -\ \ 8 \\ \hline 2\ 9 \end{array}$$

18. $68 + 94 = 162$ blocks

19. $3 \times \$5 = \$15$; $5 each

20. $6 \times 2 = 12$ pints

Systematic Review 8F

1. $4 \times 5 = 20$

2. $2 \times 2 = 4$

3. $10 \times 2 = 20$

4. $5 \times 5 = 25$

5. $9 \times 0 = 0$

6. $3 \times 10 = 30$

7. $4 \times 2 = 8$ sq mi

8. $1 \times 1 = 1$ sq in

9. $5 \times 6 = 30$ sq ft

10.
$$\begin{array}{r} 21 \\ +58 \\ \hline 79 \end{array}$$

11.
$$\begin{array}{r} {}^{1}15 \\ +17 \\ \hline 32 \end{array}$$

12.
$$\begin{array}{r} {}^{1}84 \\ +46 \\ \hline 130 \end{array}$$

Lesson Practice 9A

1. 9, 18, 27, 36, 45, 54, 63, 72, 81, 90

2. 9, 18, 27, 36, 45, 54, 63, 72, 81, 90

3. $\dfrac{2}{9}, \dfrac{4}{18}, \dfrac{6}{27}, \dfrac{8}{36}, \dfrac{10}{45}, \dfrac{12}{54}, \dfrac{14}{63}, \dfrac{16}{72}, \dfrac{18}{81}, \dfrac{20}{90}$

4. 9, 18, 27, 36, 45, 54 players

5. $9 \times 10 = 90$¢

6. 9, 18, 27 slices

Lesson Practice 9B

1. 9, 18, 27, 36, 45, 54, 63, 72, 81, 90

2. 9, 18, 27, 36, 45, 54, 63, 72, 81, 90

3. $\dfrac{5}{9}, \dfrac{10}{18}, \dfrac{15}{27}, \dfrac{20}{36}, \dfrac{25}{45}, \dfrac{30}{54}, \dfrac{35}{63}, \dfrac{40}{72}, \dfrac{45}{81}, \dfrac{50}{90}$

4. 9, 18, 27, 36 children

5. 9, 18, 27, 36, 45, 54, 63, 72 rocks

6. 9, 18, 27, 36, 45, 54, 63, 72, 81, 90 sq ft

Lesson Practice 9C

1. 9, 18, 27, 36, 45, 54, 63, 72, 81, 90
2. 9, 18, 27, 36, 45, 54, 63, 72, 81, 90
3. $\frac{9}{10}, \frac{18}{20}, \frac{27}{30}, \frac{36}{40}, \frac{45}{50}, \frac{54}{60},$
   $\frac{63}{70}, \frac{72}{80}, \frac{81}{90}, \frac{90}{100}$
4. 9, 18, 27, 36, 45, 56, 63, 72, <u>81</u> tiles
5. 9, 18, 27, 36, <u>45</u> cities
6. 9, 18, 27, 36, 45, 54, <u>63</u> letters

Systematic Review 9D

1. 9, 18, 27, 36, 45, 54, 63, 72, 81, 90
2. 5, 10, 15, 20, 25, 30, 35, 40, 45, 50
3. $5 \times \underline{5} = 25$
4. $6 \times \underline{10} = 60$
5. $2 \times \underline{9} = 18$
6. $1 \times \underline{7} = 7$
7. $5 \times \underline{4} = 20$
8. $10 \times 8 = 80$
9. $7 \times 2 = 14$ sq in
10. $10 \times 10 = 100$ sq mi
11. $5 \times 3 = 15$ sq in
12. $\begin{array}{r} 25 \\ +62 \\ \hline 87 \end{array}$
13. $\begin{array}{r} {}^8\cancel{9}\,{}^1\cancel{1} \\ -\ 4\ 5 \\ \hline 4\ 6 \end{array}$
14. $\begin{array}{r} {}^1 16 \\ +\ 17 \\ \hline 33 \end{array}$
15. $\begin{array}{r} 86 \\ -73 \\ \hline 13 \end{array}$
16. $3 \times 5 = 15$¢
17. $4 \times 5 = 20$¢
18. $10 \times 8 = 80$¢
19. $14 + 29 = 43$ arrows
20. 9, <u>18</u> players

Systematic Review 9E

1. 9, 18, 27, 36, 45, 54, 63, 72, 81, 90
2. $\frac{5}{10}, \frac{10}{20}, \frac{15}{30}, \frac{20}{40}, \frac{25}{50}, \frac{30}{60},$
   $\frac{35}{70}, \frac{40}{80}, \frac{45}{90}, \frac{50}{100}$
3. $3 \times \underline{10} = 30$
4. $4 \times \underline{2} = 8$
5. $5 \times \underline{7} = 35$
6. $0 \times 8 = 0$
7. $9 \times 10 = 90$
8. $6 \times 5 = 30$
9. $10 \times 1 = 10$ sq ft
10. $5 \times 5 = 25$ sq in
11. $5 \times 2 = 10$ sq in
12. $\begin{array}{r} {}^1 13 \\ +47 \\ \hline 60 \end{array}$
13. $\begin{array}{r} {}^6\cancel{7}\,{}^1 5 \\ -\ 1\ 8 \\ \hline 5\ 7 \end{array}$
14. $\begin{array}{r} {}^1 43 \\ +59 \\ \hline 102 \end{array}$
15. $\begin{array}{r} {}^8\cancel{9}\,{}^1 4 \\ -\ 5\ 5 \\ \hline 3\ 9 \end{array}$
16. 9, 18, 27, 36, 45, 54; 6 vans
17. $7 \times 10 = 70$¢
18. $5 \times \underline{9} = 45$; 9 rings
19. $44 + 26 = 70$
    $70 - 12 = 58$ tons
20. $6 \times 2 = 12$ jars

Systematic Review 9F

1. 9, 18, 27, 36, 45, 54, 63, 72, 81, 90
2. $\frac{2}{(2)(1)}, \frac{4}{(2)(2)}, \frac{6}{(2)(3)}, \frac{8}{(2)(4)}, \frac{10}{(2)(5)},$
   $\frac{12}{(2)(6)}, \frac{14}{(2)(7)}, \frac{16}{(2)(8)}, \frac{18}{(2)(9)}, \frac{20}{(2)(10)}$
3. $8 \times \underline{5} = 40$
4. $9 \times \underline{10} = 90$

5. $1 \times \underline{4} = 4$

6. $9 \times 1 = 9$

7. $2 \times 10 = 20$

8. $2 \times 9 = 18$

9. $3 \times 2 = 6$ sq mi

10. $2 \times 2 = 4$ sq ft

11. $9 \times 5 = 45$ sq in

12.
$$\begin{array}{r} {}^{1}\!\!\!\not{2}\,{}^{1}7 \\ -\ 1\ 9 \\ \hline 8 \end{array}$$

13.
$$\begin{array}{r} {}^{1}6\ 1 \\ +\ 3\ 9 \\ \hline 1\ 0\ 0 \end{array}$$

14.
$$\begin{array}{r} {}^{1}5\ 4 \\ +\ 4\ 7 \\ \hline 1\ 0\ 1 \end{array}$$

15.
$$\begin{array}{r} {}^{7}\!\!\not{8}\,{}^{1}2 \\ -\ 7\ 3 \\ \hline 9 \end{array}$$

16. 9, 18, 27, 36, 45, 54, 63, 72, 81; 9 days

17. $6 \times 5 = 30¢$

18. $6 \times 10 = 60¢$
    $60¢ + 30¢ = 90¢$

19. $11 + 9 = 20$ pails for Jack
    $20 - 6 = 14$ pails

20. $7 \times 10 = 70$ books

13. $9 \times 8 = 72$
    $8 \times 9 = 72$

14. $9 \times 4 = 36$
    $4 \times 9 = 36$

15. $9 \times 5 = 45$
    $5 \times 9 = 45$

16. $9 \times 7 = 63$
    $7 \times 9 = 63$

17. $9 \times 0 = 0$

18. $9 \times 10 = 90$

19. $9 \times 2 = 18$

20. $9 \times 6 = 54$

21.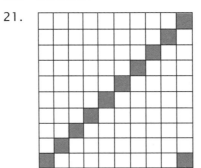

22. $9 \times 4 = 36$

Lesson Practice 10A

1. $9 \times 4 = 36$

2. $9 \times 9 = 81$

3. $9 \times 8 = 72$

4. $10 \times 9 = 90$

5. $2 \times 9 = 18$

6. $5 \times 9 = 45$

7. $9 \times 1 = 9$

8. $9 \times 3 = 27$

9. $9 \times 7 = 63$

10. $9 \times 9 = 81$

11. $6 \times 9 = 54$

12. $3 \times 9 = 27$

Lesson Practice 10B

1. $9 \times 10 = 90$

2. $9 \times 6 = 54$

3. $9 \times 2 = 18$

4. $0 \times 9 = 0$

5. $7 \times 9 = 63$

6. $3 \times 9 = 27$

7. $9 \times 5 = 45$

8. $9 \times 8 = 72$

9. $4 \times 9 = 36$

10. $1 \times 9 = 9$

11. $9 \times 9 = 81$

12. $6 \times 9 = 54$

13. $9 \times 3 = 27$
    $3 \times 9 = 27$

14. $9 \times 2 = 18$
    $2 \times 9 = 18$

15. $9 \times 6 = 54$
    $6 \times 9 = 54$
16. $9 \times 10 = 90$
    $10 \times 9 = 90$
17. $9 \times 7 = 63$
18. $9 \times 5 = 45$
19. $9 \times 8 = 72$
20. $9 \times 4 = 36$
21. $\dfrac{0}{9 \cdot 0}, \dfrac{9}{9 \cdot 1}, \dfrac{18}{9 \cdot 2}, \dfrac{27}{9 \cdot 3}, \dfrac{36}{9 \cdot 4}, \dfrac{45}{9 \cdot 5},$
    $\dfrac{54}{9 \cdot 6}, \dfrac{63}{9 \cdot 7}, \dfrac{72}{9 \cdot 8}, \dfrac{81}{9 \cdot 9}, \dfrac{90}{9 \cdot 10}$
22. $9 \times 5 = 45$ planets
23. $9 \times 9 = 81$ squares
24. $9 \times 3 = 27$ acres

Lesson Practice 10C

1. $9 \times 1 = 9$
2. $9 \times 8 = 72$
3. $9 \times 6 = 54$
4. $10 \times 2 = 20$
5. $9 \times 9 = 81$
6. $3 \times 9 = 27$
7. $5 \times 9 = 45$
8. $9 \times 7 = 63$
9. $9 \times 0 = 0$
10. $4 \times 9 = 36$
11. $10 \times 9 = 90$
12. $9 \times 8 = 72$
13. $9 \times 1 = 9$
    $1 \times 9 = 9$
14. $9 \times 7 = 63$
    $7 \times 9 = 63$
15. $9 \times 4 = 36$
    $4 \times 9 = 36$
16. $9 \times 5 = 45$
    $5 \times 9 = 45$
17. $9 \times 3 = 27$
18. $9 \times 9 = 81$
19. $9 \times 10 = 90$
20. $9 \times 0 = 0$

21.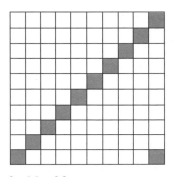

$9 \times 11 = 99$

22. $6 \times 9 = 54 ¢$

Systematic Review 10D

1. $9 \times 5 = 45$
2. $7 \times 9 = 63$
3. $5 \times 6 - 30$
4. $8 \times 2 = 16$
5. $10 \times 9 = 90$
6. $8 \times 5 = 40$
7. $3 \times 2 = 6$
8. $7 \times 5 = 35$
9. $9 \times 6 = 54$
10. $2 \times 7 - 14$
11. $10 \times 5 = 50$
12. $6 \times 0 = 0$
13. $9 \times 9 = 81$ sq in
14. $3 \times 9 = 27$ sq ft
15. $4 \times 5 = 20$ sq mi
16. $8 \times \$9 = \$72$; $\$9$ per gift
17. $9 \times 4 = 36$
    $36 + 15 = 51$ min
18. $5 \times 5 = 25$
    $3 \times 10 = 30$
    $30 + 25 = 55 ¢$
19. $35 - 28 = 7$ questions
20. $2 \times 6 = 12$
    $53 - 12 = 41$ sq in

Systematic Review10E

1. $2 \times 9 = 18$
2. $5 \times 5 = 25$
3. $7 \times 10 = 70$
4. $6 \times 9 = 54$
5. $1 \times 9 = 9$
6. $9 \times 9 = 81$
7. $5 \times 3 = 15$
8. $9 \times 5 = 45$
9. $2 \times \underline{10} = 20$
10. $7 \times \underline{9} = 63$
11. $2 \times \underline{5} = 10$
12. $3 \times \underline{9} = 27$
13. $9 \times 4 = 36$ sq in
14. $2 \times 7 = 14$ sq ft
15. $1 \times 1 = 1$ sq in
16. $\dfrac{0}{9 \times 0}, \dfrac{9}{9 \times 1}, \dfrac{18}{9 \times 2}, \dfrac{27}{9 \times 3}, \dfrac{36}{9 \times 4}, \dfrac{45}{9 \times 5},$
    $\dfrac{54}{9 \times 6}, \dfrac{63}{9 \times 7}, \dfrac{72}{9 \times 8}, \dfrac{81}{9 \times 9}, \dfrac{90}{9 \times 10}$
17. $6 \times 10 = 60$¢
    $7 \times 5 = 35$¢
    $60 + 35 = 95$¢
18. $9 \times 8 = 72$ sq mi
19. $13 + 12 = 25$
    $25 - 7 = 18$ pints
20. $45 + 39 = 84$ miles

13. $9 \times 7 = 63$ sq in
14. $10 \times 10 = 100$ sq ft
15. $2 \times 6 = 12$ sq ft
16. $7 \times 5 = 35$¢ - Jim
    $4 \times 10 = 40$¢ - Lisa
    Lisa has more
    $35 + 40 = 75$¢
17. $9 \times 9 = 81$
    $81 - 3 = 78$ clubs
18. $17 + 25 = 42$ children
19. $53 - 45 = 8$ minutes
20. $5 \times 9 = 45$ tiles

Lesson Practice 11A

1. 3, 6, 9, 12, 15, 18, 21, 24, 27, 30
2. 3, 6, 9, 12, 15, 18, 21, 24, 27, 30
3. 3, 6, 9, 12, 15, 18, 21, 24, 27, 30
4. $\dfrac{2}{3} = \dfrac{4}{6} = \dfrac{6}{9} = \dfrac{8}{12} = \dfrac{10}{15} = \dfrac{12}{18} =$
   $\dfrac{14}{21} = \dfrac{16}{24} = \dfrac{18}{27} = \dfrac{20}{30}$
5. 3, 6, 9, 12, $\underline{15}$ wheels
6. 3, 6, 9, 12, 15, $\underline{18}$ years
7. 3, 6, 9, 12, 15, 18, 21, $\underline{24}$ sides
8. 3, 6, 9, $\underline{12}$ dots

Systematic Review10F

1. $2 \times 4 = 8$
2. $5 \times 6 = 30$
3. $9 \times 5 = 45$
4. $3 \times 9 = 27$
5. $8 \times 9 = 72$
6. $9 \times 6 = 54$
7. $10 \times 3 = 30$
8. $4 \times 5 = 20$
9. $4 \times \underline{9} = 36$
10. $2 \times \underline{9} = 18$
11. $8 \times \underline{10} = 80$
12. $8 \times \underline{5} = 40$

Lesson Practice 11B

1. 3, 6, 9, 12, 15, 18, 21, 24, 27, 30
2. 3, 6, 9, 12, 15, 18, 21, 24, 27, 30
3. 3, 6, 9, 12, 15, 18, 21, 24, 27, 30
4. $\dfrac{3}{5} = \dfrac{6}{10} = \dfrac{9}{15} = \dfrac{12}{20} = \dfrac{15}{25} = \dfrac{18}{30} =$
   $\dfrac{21}{35} = \dfrac{24}{40} = \dfrac{27}{45} = \dfrac{30}{50}$
5. 3, 6, $\underline{9}$ sides
6. 3, 6, 9, 12, 15, 18, 21, 24, $\underline{27}$ baby steps
7. 3, 6, 9, 12, 15, 18, 21, $\underline{24}$ people
8. 3, $\underline{6}$ hands

Lesson Practice 11C

1. 3, 6, 9, 12, 15, 18, 21, 24, 27, 30
2. 3, 6, 9, 12, 15, 18, 21, 24, 27, 30
3. 3, 6, 9, 12, 15, 18, 21, 24, 27, 30
4. $\frac{3}{10} = \frac{6}{20} = \frac{9}{30} = \frac{12}{40} = \frac{15}{50} =$
   $\frac{18}{60} = \frac{21}{70} = \frac{24}{80} = \frac{27}{90} = \frac{30}{100}$
5. 3, 6, 9, <u>12</u> shirts
6. 3, 6, 9, 12, 15, 18, <u>21</u> books
7. 3, 6, 9, 12, 15, 18, 21, 24, 27, <u>$30</u>
8. 3, <u>6</u> things

Systematic Review 11D

1. 3, 6, 9, 12, 15, 18, 21, 24, 27, 30
2. $\frac{0}{(9)(0)}, \frac{9}{(9)(1)}, \frac{18}{(9)(2)}, \frac{27}{(9)(3)},$
   $\frac{36}{(9)(4)}, \frac{45}{(9)(5)}, \frac{54}{(9)(6)}, \frac{63}{(9)(7)},$
   $\frac{72}{(9)(8)}, \frac{81}{(9)(9)}, \frac{90}{(9)(10)}$
3. $9 \times \underline{3} = 27$
4. $5 \times 4 = 20$
5. $2 \times \underline{0} = 0$
6. $10 \times \underline{5} = 50$
7. $2 \times 8 = 16$
8. $1 \times 7 = 7$
9. $7 \times 10 = 70$
10. $9 \times 4 = 36$
11. $9 \times 6 = 54$ sq in
12. $5 \times 5 = 25$ sq mi
13. $7 \times 5 = 35$ sq in
14. $\begin{array}{r} {}^1 2\,3 \\ 2\,6 \\ +\,3\,7 \\ \hline 8\,6 \end{array}$
15. $\begin{array}{r} {}^1 1\,2 \\ 5\,9 \\ +\,3\,1 \\ \hline 1\,0\,2 \end{array}$
16. $\begin{array}{r} {}^1 1\,5 \\ 1\,5 \\ 4\,4 \\ +\,2\,4 \\ \hline 9\,8 \end{array}$
17. $\begin{array}{r} {}^2 3\,4 \\ 5\,6 \\ 1\,1 \\ +\,\,\,9 \\ \hline 1\,1\,0 \end{array}$
18. $3+5+4+5+6+2 = 25$ pies

Systematic Review 11E

1. 3, 6, 9, 12, 15, 18, 21, 24, 27, 30
2. $\frac{0}{5 \cdot 0}, \frac{5}{5 \cdot 1}, \frac{10}{5 \cdot 2}, \frac{15}{5 \cdot 3}, \frac{20}{5 \cdot 4}, \frac{25}{5 \cdot 5},$
   $\frac{30}{5 \cdot 6}, \frac{35}{5 \cdot 7}, \frac{40}{5 \cdot 8}, \frac{45}{5 \cdot 9}, \frac{50}{5 \cdot 10}$
3. $7 \times \underline{9} = 63$
4. $6 \times \underline{5} = 30$
5. $2 \times \underline{4} = 8$
6. $6 \times \underline{10} = 60$
7. $3 \times 5 = 15$
8. $8 \times 9 = 72$
9. $9 \times 5 = 45$
10. $8 \times 1 = 8$
11. $8 \times 5 = 40$ sq in
12. $2 \times 2 = 4$ sq mi
13. $9 \times 3 = 27$ sq in
14. $\begin{array}{r} {}^1 4\,5 \\ 3\,1 \\ +\,1\,5 \\ \hline 9\,1 \end{array}$
15. $\begin{array}{r} {}^1 2\,6 \\ 8\,4 \\ +\,3\,2 \\ \hline 1\,4\,2 \end{array}$
16. $\begin{array}{r} {}^1 2\,3 \\ 1\,5 \\ 1\,7 \\ +\,\,\,2 \\ \hline 5\,7 \end{array}$

17.
$$
\begin{array}{r}
{}^{2}3\,1 \\
2\,9 \\
3\,2 \\
+8 \\
\hline
1\,0\,0
\end{array}
$$

18. $9 \times 10 = 90¢$

$2 \times 5 = 10¢$

$90 + 10 = 100¢ = \$1.00$

19. 3, 6, 9, 12, 15, $\underline{18}$ sides

20. $52 - 35 = \underline{17}$ minutes

17.
$$
\begin{array}{r}
{}^{1}5\,1 \\
1\,2 \\
2\,4 \\
+3\,8 \\
\hline
1\,2\,5
\end{array}
$$

18. $5 \times \underline{9} = 45¢$; 9 nickels

19. 3, 6, 9, 12, 15, 18, $\underline{21}$ dots

20. $\$50 - \$38 = \$12$

$3 \times \underline{4} = \$12$; \$4 each

Systematic Review 11F

1. 3, 6, 9, 12, 15, 18, 21, 24, 27, 30

2. $\dfrac{0}{2 \times 0}, \dfrac{2}{2 \times 1}, \dfrac{4}{2 \times 2}, \dfrac{6}{2 \times 3}, \dfrac{8}{2 \times 4}, \dfrac{10}{2 \times 5},$

$\dfrac{12}{2 \times 6}, \dfrac{14}{2 \times 7}, \dfrac{16}{2 \times 8}, \dfrac{18}{2 \times 9}, \dfrac{20}{2 \times 10}$

3. $2 \times \underline{3} = 6$

4. $4 \times \underline{10} = 40$

5. $6 \times \underline{9} = 54$

6. $5 \times \underline{5} = 25$

7. $9 \times 9 = 81$

8. $4 \times 9 = 36$

9. $5 \times 7 = 35$

10. $6 \times 0 = 0$

11. $9 \times 7 = 63$ sq ft

12. $1 \times 1 = 1$ sq ft

13. $6 \times 2 = 12$ sq in

14.
$$
\begin{array}{r}
{}^{1}1\,9 \\
9\,1 \\
+7 \\
\hline
1\,1\,7
\end{array}
$$

15.
$$
\begin{array}{r}
{}^{1}1\,7 \\
3\,6 \\
+4\,4 \\
\hline
9\,7
\end{array}
$$

16.
$$
\begin{array}{r}
{}^{1}5\,5 \\
4\,1 \\
6\,5 \\
+2 \\
\hline
1\,6\,3
\end{array}
$$

Lesson Practice 12A

1. $3 \times 4 = 12$

2. $9 \times 3 = 27$

3. $3 \times 7 = 21$

4. $10 \times 3 = 30$

5. $3 \times 7 = 21$

6. $1 \times 3 = 3$

7. $3 \times 4 = 12$

8. $8 \times 3 = 24$

9. $8 \times 3 = 24$

10. $3 \times 5 = 15$

11. $3 \times 3 = 9$

12. $6 \times 3 = 18$

13. 3, $\underline{6}$

14. 3, 6, 9, 12, 15, $\underline{18}$

15. (See lesson 12 in instruction manual.)

16. $\dfrac{0}{(3)(0)}, \dfrac{3}{(3)(1)}, \dfrac{6}{(3)(2)}, \dfrac{9}{(3)(3)},$

$\dfrac{12}{(3)(4)}, \dfrac{15}{(3)(5)}, \dfrac{18}{(3)(6)}, \dfrac{21}{(3)(7)},$

$\dfrac{24}{(3)(8)}, \dfrac{27}{(3)(9)}, \dfrac{30}{(3)(10)}$

17. $\dfrac{3}{5} = \dfrac{6}{10} = \dfrac{9}{15} = \dfrac{12}{20} = \dfrac{15}{25} =$

$\dfrac{18}{30} = \dfrac{21}{35} = \dfrac{24}{40} = \dfrac{27}{45} = \dfrac{30}{50}$

18. $4 \times 3 = 12$ ft

19. $6 \times 3 = 18$ tsp

20. $3 \times 3 = 9$ ft

Lesson Practice 12B

1. $3 \times 10 = 30$
2. $3 \times 3 = 9$
3. $3 \times 5 = 15$
4. $6 \times 3 = 18$
5. $3 \times 2 = 6$
6. $0 \times 3 = 0$
7. $3 \times 9 = 27$
8. $7 \times 3 = 21$
9. $1 \times 3 = 3$
10. $3 \times 8 = 24$
11. $4 \times 3 = 12$
12. $10 \times 3 = 30$
13. 3, 6, 9, 12, 15, 18, 21, 24, <u>27</u>
14. 3, 6, 9, 12, <u>15</u>
15. $3 \times 6 = 18$
    $6 \times 3 = 18$
16. $3 \times 2 = 6$
    $2 \times 3 = 6$
17. $3 \times 8 = 24'$
18. $7 \times 3 = 21$
19. $4 \times 3 = 12$
20. $5 \times 3 = 15$ pillows

Lesson Practice 12C

1. $3 \times 1 = 3$
2. $3 \times 5 = 15$
3. $3 \times 10 = 30$
4. $2 \times 3 = 6$
5. $3 \times 3 = 9$
6. $9 \times 3 = 27$
7. $3 \times 6 = 18$
8. $4 \times 3 = 12$
9. $7 \times 3 = 21$
10. $3 \times 0 = 0$
11. $8 \times 3 = 24$
12. $5 \times 3 = 15$
13. 3, 6, 9, 12, 15, 18, 21, <u>24</u>
14. 3, 6, <u>9</u>

15.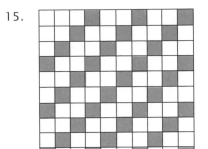
    $3 \times 12 = 36$
16. $3 \times 2 = 6'$
17. $9 \times 3 = 27$
18. $4 + 1 + 1 = 6$
    $6 \times 3 = 18$ tsp

Systematic Review 12D

1. $3 \times 3 = 9$
2. $8 \times 3 = 24$
3. $3 \times 7 = 21$
4. $6 \times 3 = 18$
5. $5 \times 9 = 45$
6. $6 \times 2 = 12$
7. $5 \times 5 = 25$
8. $8 \times 9 = 72$
9. $4 \times 2 = 8$
10. $7 \times 5 = 35¢$
11. $5 \times 3 = 15$
12. $6 \times 3 = 18$
13.
$$\begin{array}{r} {}^{1}3\,1 \\ 7\,9 \\ 4\,5 \\ +\ \ 3 \\ \hline 1\,5\,8 \end{array}$$
14.
$$\begin{array}{r} {}^{1}1\,8 \\ 2\,5 \\ 5\,3 \\ +7\,2 \\ \hline 1\,6\,8 \end{array}$$
15.
$$\begin{array}{r} 6\,9 \\ -1\,9 \\ \hline 5\,0 \end{array}$$

16.
$$\begin{array}{r} {}^7\!\!\!\!8\,{}^1\!1 \\ -\ 2\ 7 \\ \hline 5\ 4 \end{array}$$

17. $3 \times \underline{4} = 12$; 4 yards

18. $3 \times 9 = 27$ children

19. $5 + 13 + 7 + 15 = 40$ leaves

20. $\$8 + \$18 = \$26$
$\$26 - \$7 = \$19$

Systematic Review 12E

1. $3 \times 5 = 15$
2. $10 \times 3 = 30$
3. $3 \times 6 = 18$
4. $8 \times 3 = 24$
5. $5 \times 8 = 40$
6. $7 \times 9 = 63$
7. $10 \times 6 = 60$
8. $8 \times 2 = 16$
9. $7 \times 2 = 14$
10. $4 \times 10 = 40$
11. $9 \times 3 = 27$
12. $3 \times 3 = 9$

13.
$$\begin{array}{r} {}^1\!5\ 7 \\ 2\ 1 \\ 2\ 2 \\ +1\ 5 \\ \hline 1\ 1\ 5 \end{array}$$

14.
$$\begin{array}{r} {}^1\!4\ 3 \\ 4\ 4 \\ 6\ 3 \\ +1\ 1 \\ \hline 1\ 6\ 1 \end{array}$$

15.
$$\begin{array}{r} {}^4\!\!\!\!5\,{}^1\!3 \\ -\ 2\ 4 \\ \hline 2\ 9 \end{array}$$

16.
$$\begin{array}{r} {}^5\!\!\!\!6\,{}^1\!5 \\ -\ 1\ 9 \\ \hline 4\ 6 \end{array}$$

17. $2 \times 3 = 6$ tsp
$\underline{2\ \text{Tbsp} \times 3 = 6\ \text{tsp}}$

18. $3 \times 3 = 9$ apples
$3 \times 9 = 27$ worms

19. $12 + 25 + 3 + 10 = 50$ miles

20. $3 \times 7 = 21$ sq yd
$21 - 11 = 10$ sq yd

Systematic Review 12F

1. $4 \times 3 = 12$
2. $7 \times 3 = 21$
3. $3 \times 8 = 24$
4. $1 \times 3 = 3$
5. $10 \times 3 = 30$
6. $6 \times 3 = 18$
7. $9 \times 6 = 54$
8. $5 \times 4 = 20$
9. $9 \times 2 = 18$
10. $6 \times 5 = 30$
11. $3 \times 3 = 9$
12. $5 \times 3 = 15$

13.
$$\begin{array}{r} {}^1\!1\ 7 \\ 1\ 6 \\ 1\ 2 \\ +1\ 4 \\ \hline 5\ 9 \end{array}$$

14.
$$\begin{array}{r} {}^1\!3\ 5 \\ 7\ 4 \\ 1\ 5 \\ +2\ 4 \\ \hline 1\ 4\ 8 \end{array}$$

15.
$$\begin{array}{r} {}^3\!\!\!\!4\,{}^1\!1 \\ -\ 3\ 2 \\ \hline 9 \end{array}$$

16.
$$\begin{array}{r} {}^2\!\!\!\!3\,{}^1\!8 \\ -\ 2\ 9 \\ \hline 9 \end{array}$$

17. $3 \times \underline{10} = 30$; 10 yards

18. $3 \times 9 = 27$
$30 - 27 = \$3$

19. $57 + 14 = 71$
$71 - 71 = 0$ dogs

20. $4 \times 10 = 40$¢

$3 \times 5 = 15$¢

$15 + 40 + 11 = 66$¢

Lesson Practice 13A

1. 6, 12, 18, 24, 30, 36, 42, 48, 54, 60
2. 6, 12, 18, 24, 30, 36, 42, 48, 54, 60
3. 6, 12, 18, 24, 30, 36, 42, 48, 54, 60
4. $\frac{1}{2} = \frac{2}{4} = \frac{3}{6} = \frac{4}{8} = \frac{5}{10}$
5. $\frac{1}{3} = \frac{2}{6} = \frac{3}{9} = \frac{4}{12} = \frac{5}{15}$
6. 6, 12, 18 paintings
7. 6, 12, 18, 24, 30, 36, 42, 48 crackers

Lesson Practice 13B

1. 6, 12, 18, 24, 30, 36, 42, 48, 54, 60
2. 6, 12, 18, 24, 30, 36, 42, 48, 54, 60
3. 6, 12, 18, 24, 30, 36, 42, 48, 54, 60
4. $\frac{1}{6} = \frac{2}{12} = \frac{3}{18} = \frac{4}{24} = \frac{5}{30}$
5. $\frac{2}{6} = \frac{4}{12} = \frac{6}{18} = \frac{8}{24} = \frac{10}{30}$
6. 6, 12 cones
7. 6, 12, 18, 24, 30, 36 people

Lesson Practice 13C

1. 6, 12, 18, 24, 30, 36, 42, 48, 54, 60
2. 6, 12, 18, 24, 30, 36, 42, 48, 54, 60
3. 6, 12, 18, 24, 30, 36, 42, 48, 54, 60
4. $\frac{5}{6} = \frac{10}{12} = \frac{15}{18} = \frac{20}{24} = \frac{25}{30}$
5. $\frac{2}{3} = \frac{4}{6} = \frac{6}{9} = \frac{8}{12} = \frac{10}{15}$
6. 6, 12, 18, 24, 30, 36, 42, 48, 54 times
7. 6, 12, 18, 24, 30 fingers

Systematic Review 13D

1. 6, 12, 18, 24, 30, 36, 42, 48, 54, 60
2. $\frac{3}{5} = \frac{6}{10} = \frac{9}{15} = \frac{12}{20} = \frac{15}{25}$
3. $9 \times 3 = 27$
4. $2 \times 6 = 12$
5. $3 \times 4 = 12$
6. $5 \times 5 = 25$
7. $8 \times 9 = 72$
8. $6 \times 5 = 30$
9. $3 \times 7 = 21$
10. $10 \times 8 = 80$
11. $\begin{array}{r} 73 \\ +45 \\ \hline 118 \end{array}$
12. $\begin{array}{r} {}^{1}38 \\ + 67 \\ \hline 1\,05 \end{array}$
13. $\begin{array}{r} {}^{4}\!5\,{}^{1}4 \\ - \ 2\ 5 \\ \hline 2\ 9 \end{array}$
14. $\begin{array}{r} {}^{7}\!8\,{}^{1}8 \\ - \ 1\ 9 \\ \hline 6\ 9 \end{array}$
15. done
16. $3 + 3 + 3 + 3 = 12$ mi
17. $10 + 12 + 10 + 12 = 44'$
18. $25 + 25 + 25 + 25 = 100'$
19. 6, 12, 18, 24, 30, 36 eggs
20. 6, 12, 18 ft

Systematic Review 13E

1. 6, 12, 18, 24, 30, 36, 42, 48, 54, 60
2. $\frac{2}{5} = \frac{4}{10} = \frac{6}{15} = \frac{8}{20} = \frac{10}{25}$
3. $2 \times 7 = 14$
4. $3 \times 0 = 0$
5. $9 \times 9 = 81$
6. $5 \times 3 = 15$
7. $3 \times 3 = 9$
8. $9 \times 4 = 36$
9. $10 \times 7 = 70$

10. $8 \times 2 = 16$

11.
$$\begin{array}{r} 61 \\ +22 \\ \hline 83 \end{array}$$

12.
$$\begin{array}{r} {}^{1}45 \\ +9 \\ \hline 54 \end{array}$$

13.
$$\begin{array}{r} {}^{6}\cancel{7}\,{}^{1}6 \\ -38 \\ \hline 38 \end{array}$$

14.
$$\begin{array}{r} {}^{8}\cancel{9}\,{}^{1}3 \\ -44 \\ \hline 49 \end{array}$$

15. $16 + 20 + 16 + 20 = 72$ yd

16. $7 + 4 + 7 + 4 = 22'$

17. $5 + 5 + 5 + 5 = 20''$

18. $6 + 4 + 6 + 4 = 20$ mi

19. 6, 12, 18, 24, 30, 36, 40, <u>48</u> ft

20. $8 \times 3 = 24$ tsp

Systematic Review 13F

1. 9, 18, 27, 36, 45, 54, 63, 72, 81, 90

2. $\dfrac{3}{6} = \dfrac{6}{12} = \dfrac{9}{18} = \dfrac{12}{24} = \dfrac{15}{30}$

3. $3 \times 9 = 27$

4. $9 \times 2 = 18$

5. $2 \times 4 = 8$

6. $10 \times 6 = 60$

7. $9 \times 5 = 45$

8. $5 \times 4 = 20$

9. $9 \times 6 = 54$

10. $7 \times 9 = 63$

11.
$$\begin{array}{r} {}^{1}17 \\ +18 \\ \hline 35 \end{array}$$

12.
$$\begin{array}{r} {}^{1}34 \\ +79 \\ \hline 113 \end{array}$$

13.
$$\begin{array}{r} 48 \\ -27 \\ \hline 21 \end{array}$$

14.
$$\begin{array}{r} {}^{4}\cancel{5}\,{}^{1}3 \\ -19 \\ \hline 34 \end{array}$$

15. $11 + 14 + 11 + 14 = 50$ mi

16. $12 + 21 + 12 + 21 = 66'$

17. $10 + 10 + 10 + 10 = 40''$

18. $5 + 1 + 5 + 1 = 12''$

19. $5 \times 1 = 5$ sq in

20. $5 \times 2 = 10$

$10 \times 9 = 90$ fingers

Lesson Practice 14A

1. $6 \times 9 = 54$

2. $7 \times 6 = 42$

3. $10 \times 6 = 60$

4. $4 \times 6 = 24$

5. $5 \times 6 = 30$

6. $1 \times 6 = 6$

7. $6 \times 3 = 18$

8. $6 \times 2 = 12$

9. $9 \times 6 = 54$

10. $6 \times 4 = 24$

11. $6 \times 8 = 48$

12. $7 \times 6 = 42$

13. $6 \times 4 = 24$

14. $6 \times 1 = 6$

15. $6 \times 5 = 30$

16. $6 \times 3 = 18$

17. 6, 12, 18, 24, 30, 36, 42, <u>48</u>

18. 6, 12, 18, 24, 30, <u>36</u>

19.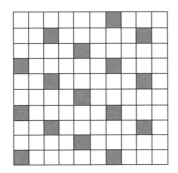

$6 \times 11 = 66$

$6 \times 12 = 72$

20. $6 \times 7 = 42$ legs

Lesson Practice 14B

1. $6 \times 0 = 0$
2. $6 \times 6 = 36$
3. $2 \times 6 = 12$
4. $8 \times 6 = 48$
5. $3 \times 6 = 18$
6. $5 \times 6 = 30$
7. $6 \times 9 = 54$
8. $6 \times 10 = 60$
9. $6 \times 6 = 36$
10. $6 \times 1 = 6$
11. $6 \times 4 = 24$
12. $2 \times 6 = 12$
13. $6 \times 7 = 42$
14. $6 \times 5 = 30$
15. $6 \times 3 = 18$
16. $9 \times 6 = 54$
17. (See lesson 14 in instruction manual.)
18. $\dfrac{0}{6 \cdot 0}, \dfrac{6}{6 \cdot 1}, \dfrac{12}{6 \cdot 2}, \dfrac{18}{6 \cdot 3}, \dfrac{24}{6 \cdot 4}, \dfrac{30}{6 \cdot 5},$
   $\dfrac{36}{6 \cdot 6}, \dfrac{42}{6 \cdot 7}, \dfrac{48}{6 \cdot 8}, \dfrac{54}{6 \cdot 9}, \dfrac{60}{6 \cdot 10}$
19. $6 \times 10 = 60$
20. $6 \times 4 = 24$ wheels

Lesson Practice 14C

1. $6 \times 10 = 60$
2. $6 \times 8 = 48$
3. $6 \times 6 = 36$
4. $4 \times 6 = 24$
5. $7 \times 6 = 42$
6. $9 \times 6 = 54$
7. $6 \times 5 = 30$
8. $6 \times 3 = 18$
9. $6 \times 0 = 0$
10. $6 \times 6 = 36$
11. $6 \times 2 = 12$
12. $8 \times 6 = 48$
13. $6 \times 5 = 30$
14. $4 \times 6 = 24$
15. $6 \times 1 = 6$
16. $7 \times 6 = 42$
17. $\dfrac{0}{6 \times 0}, \dfrac{6}{6 \times 1}, \dfrac{12}{6 \times 2}, \dfrac{18}{6 \times 3}, \dfrac{24}{6 \times 4}, \dfrac{30}{6 \times 5},$
   $\dfrac{36}{6 \times 6}, \dfrac{42}{6 \times 7}, \dfrac{48}{6 \times 8}, \dfrac{54}{6 \times 9}, \dfrac{60}{6 \times 10}$
18. $6 \times 9 = 54$
19. $6 \times 6 = 36$ feet
20. $6 \times 6 = 36$ apartments

Systematic Review 14D

1. $10 \times 7 = 70$
2. $9 \times 6 = 54$
3. $5 \times 0 = 0$
4. $6 \times 3 = 18$
5. $3 \times 9 = 27$
6. $5 \times 4 = 20$
7. $4 \times 6 = 24$
8. $7 \times 9 = 63$
9. $6 \times 5 = 30$
10. $6 \times 6 = 36$
11. $9 \times 5 = 45$
12. $7 \times 2 = 14$
13. $14 + 2 + 14 + 2 = 32"$
14. $8 + 10 + 8 = 26"$
15. $7 + 7 + 7 + 7 = 28'$

16. $\dfrac{1}{6} = \dfrac{2}{12} = \dfrac{3}{18} = \dfrac{4}{24} = \dfrac{5}{30}$

17. $6 \times 7 = 42$ glasses

18. $5 \times 10 = 50$

$3 \times 5 = 15$

$50 + 15 + 8 = 73¢$

19. $50 - 23 = 27$ apples

20. $4 + 8 = 12$

$12 - 6 = 6$ people

Systematic Review 14E

1. $3 \times 9 = 27$
2. $6 \times 9 = 54$
3. $5 \times 5 = 25$
4. $9 \times 8 = 72$
5. $4 \times 3 = 12$
6. $6 \times 8 = 48$
7. $9 \times 0 = 0$
8. $7 \times 5 = 35$
9. $3 \times 6 = 18$
10. $9 \times 9 = 81$
11. $1 \times 8 = 8$
12. $6 \times 7 = 42$
13. $7 + 9 + 7 + 9 = 32"$
14. $5 + 6 + 3 = 14'$
15. $9 + 9 + 9 + 9 = 36$ mi
16. $\dfrac{3}{5} = \dfrac{6}{10} = \dfrac{9}{15} = \dfrac{12}{20} = \dfrac{15}{25}$
17. $8 \times 6 = 48$ nails
18. $4 + 16 + 7 + 19 + 6 = 52$ cards
19. $10 \times 3 = 30'$
20. $21 + 30 + 21 + 30 = 102'$

Systematic Review 14F

1. $4 \times 9 = 36$
2. $3 \times 3 = 9$
3. $6 \times 6 = 36$
4. $7 \times 6 = 42$
5. $9 \times 7 = 63$
6. $3 \times 8 = 24$

7. $9 \times 5 = 45$
8. $4 \times 6 = 24$
9. $8 \times 2 = 16$
10. $9 \times 3 = 27$
11. $6 \times 3 = 18$
12. $6 \times 5 = 30$
13. $17 + 4 + 17 + 4 = 42"$
14. $7 + 8 + 8 = 23'$
15. $4 + 4 + 4 + 4 = 16$ yd
16. $\dfrac{2}{3} = \dfrac{4}{6} = \dfrac{6}{9} = \dfrac{8}{12} = \dfrac{10}{15}$
17. $5 \times 3 = 15$ tsp
18. $4 \times 5¢ = 20¢$; 4 nickels
19. $3 \times 4 = 12$

$7 \times 2 = 14$

$12 + 14 = 26$ feet

20. $3 \times 6 = 18$ sq ft

Lesson Practice 15A

1. 4, 8, 12, 16, 20, 24, 28, 32, 36, 40
2. 4, 8, 12, 16, 20, 24, 28, 32, 36, 40
3. 4, 8, 12, 16, 20, 24, 28, 32, 36, 40
4. $\dfrac{4}{5} = \dfrac{8}{10} = \dfrac{12}{15} = \dfrac{16}{20} = \dfrac{20}{25} =$

$\dfrac{24}{30} = \dfrac{28}{35} = \dfrac{32}{40} = \dfrac{36}{45} = \dfrac{40}{50}$

5. 4, 8, 12, 16, 20, 24, 28, <u>32</u> qt
6. 4, 8, <u>12</u>

$12 + 12 = 24$ socks

7. 4, 8, 12, 16, 20, 24, 28, 32, <u>36</u> people
8. 4, 8, 12, 16, <u>20</u> jugs

Lesson Practice 15B

1. 4, 8, 12, 16, 20, 24, 28, 32, 36, 40
2. 4, 8, 12, 16, 20, 24, 28, 32, 36, 40
3. 4, 8, 12, 16, 20, 24, 28, 32, 36, 40
4. $\dfrac{4}{9} = \dfrac{8}{18} = \dfrac{12}{27} = \dfrac{16}{36} = \dfrac{20}{45} =$

$\dfrac{24}{54} = \dfrac{28}{63} = \dfrac{32}{72} = \dfrac{36}{81} = \dfrac{40}{90}$

5. 4, 8, 12, 16, 20, 24, 28, 32, 36, <u>40</u> qt

6. 4, 8, 12, <u>16</u> tires

7. 4, 8, 12, 16, 20, <u>24</u> laps

8. 4, <u>8</u> qt

19. 4, <u>8</u> rings

20. 4, 8, 12, <u>16</u> qt

Lesson Practice 15C

1. 4, 8, 12, 16, 20, 24, 28, 32, 36, 40

2. 4, 8, 12, 16, 20, 24, 28, 32, 36, 40

3. 4, 8, 12, 16, 20, 24, 28, 32, 36, 40

4. $\dfrac{2}{4} = \dfrac{4}{8} = \dfrac{6}{12} = \dfrac{8}{16} = \dfrac{10}{20} =$
$\dfrac{12}{24} = \dfrac{14}{28} = \dfrac{16}{32} = \dfrac{18}{36} = \dfrac{20}{40}$

5. 4, 8, 12, 16, 20, 24, <u>28</u> qt

6. 4, 8, <u>12</u> pictures

7. 4, 8, 12, 16, 20, 24, 28, <u>32</u> flowers

8. 4, 8, 12, 16, 20, 24, 28, 32, <u>36</u> jars

Systematic Review 15D

1. 4, 8, 12, 16, 20, 24, 28, 32, 36, 40

2. $\dfrac{3}{4}, \dfrac{6}{8}, \dfrac{9}{12}, \dfrac{12}{16}, \dfrac{15}{20}$

3. $10 \times 7 = 70$

4. $9 \times 6 = 54$

5. $5 \times 0 = 0$

6. $6 \times 3 = 18$

7. $9 \times 3 = 27$

8. $5 \times 4 = 20$

9. $6 \times 4 = 24$

10. $7 \times 9 = 63$

11. $5 \times \underline{6} = 30$

12. $6 \times \underline{3} = 18$

13. $9 \times \underline{5} = 45$

14. $6 \times \underline{6} = 36$

15. done

16. done

17. $10 \times 10 = 100$ sq ft

18. $10 + 10 + 10 + 10 = 40$ ft

Systematic Review 15E

1. 3, 6, 9, 12, 15, 18, 21, 24, 27, 30

2. $\dfrac{1}{4} = \dfrac{2}{8} = \dfrac{3}{12} = \dfrac{4}{16} = \dfrac{5}{20}$

3. $2 \times 7 = 14$

4. $9 \times 3 = 27$

5. $9 \times 6 = 54$

6. $5 \times 5 = 25$

7. $9 \times 8 = 72$

8. $4 \times 3 = 12$

9. $6 \times 8 = 48$

10. $9 \times 0 = 0$

11. $9 \times \underline{0} = 0$

12. $7 \times \underline{5} = 35$

13. $8 \times \underline{9} = 72$

14. $9 \times \underline{3} = 27$

15. $3 \times 7 = 21$ sq ft

16. $3 + 7 + 3 + 7 = 20$ ft

17. $9 \times 9 = 81$ sq mi

18. $9 + 9 + 9 + 9 = 36$ mi

19. 6, 12, 18, <u>24</u> feet

20. $3 \times 4 = 12$ ft
$3 \times 6 = 18$ ft
$3 \times 9 = 27$ ft
$12 + 18 + 27 = 57$ ft

Systematic Review 15F

1. 4, 8, 12, 16, 20, 24, 28, 32, 36, 40

2. $\dfrac{2}{5} = \dfrac{4}{10} = \dfrac{6}{15} = \dfrac{8}{20} = \dfrac{10}{25}$

3. $8 \times 1 = 8$

4. $7 \times 6 = 42$

5. $4 \times 9 = 36$

6. $3 \times 3 = 9$

7. $6 \times 6 = 36$

8. $7 \times 6 = 42$

9. $7 \times 9 = 63$
10. $3 \times 8 = 24$
11. $3 \times \underline{6} = 18$
12. $6 \times \underline{4} = 24$
13. $2 \times \underline{8} = 16$
14. $9 \times \underline{9} = 81$
15. $5 \times 6 = 30$ sq ft
16. $5 + 6 + 5 + 6 = 22$ ft
17. $2 \times 10 = 20$ sq yd
18. $2 + 10 + 2 + 10 = 24$ yd
19. 4, 8, 12, 16, 20, 24, $\underline{28}$ days
20. $2 \times 3 = 6$
    $2 \times 7 = 14$
    $2 \times 2 = 4$
    $2 \times 5 = 10$
    $6 + 14 + 4 + 10 = 34$ pints

Lesson Practice 16A

1. $4 \times 9 = 36$
2. $7 \times 4 = 28$
3. $10 \times 4 = 40$
4. $4 \times 6 = 24$
5. $5 \times 4 = 20$
6. $1 \times 4 = 4$
7. $4 \times 3 = 12$
8. $4 \times 2 = 8$
9. $4 \times 4 = 16$
10. $4 \times 8 = 32$
11. $4 \times 5 = 20$
12. $6 \times 4 = 24$
13. $4 \times 9 = 36$
14. $4 \times 4 = 16$
15.

$4 \times 11 = 44$
$4 \times 12 = 48$
16. $4 \times 6 = 24$ quarters
17. $7 \times 4 = 28$ corners
18. $3 \times 4 = 12$ quarters

Lesson Practice 16B

1. $4 \times 0 = 0$
2. $4 \times 10 = 40$
3. $4 \times 3 = 12$
4. $6 \times 4 = 24$
5. $2 \times 4 = 8$
6. $4 \times 4 = 16$
7. $4 \times 7 = 28$
8. $9 \times 4 = 36$
9. $1 \times 4 = 4$
10. $4 \times 5 = 20$
11. $4 \times 8 = 32$
12. $3 \times 4 = 12$
13. 4, 8, 12, 16, 20, 24, 28, 32, 36, 40
14. $\dfrac{0}{(4)(0)}, \dfrac{4}{(4)(1)}, \dfrac{8}{(4)(2)}, \dfrac{12}{(4)(3)}, \dfrac{16}{(4)(4)}, \dfrac{20}{(4)(5)},$

    $\dfrac{24}{(4)(6)}, \dfrac{28}{(4)(7)}, \dfrac{32}{(4)(8)}, \dfrac{36}{(4)(9)}, \dfrac{40}{(4)(10)}$
15. $4 \times 4 = 16$ quarters
16. $4 \times 7 = 28$
17. $4 \times 9 = 36$ plates
18. $4 \times 5 = 20$ horseshoes

Lesson Practice 16C

1. $4 \times 2 = 8$
2. $4 \times 4 = 16$
3. $6 \times 4 = 24$
4. $4 \times 10 = 40$
5. $1 \times 4 = 4$
6. $4 \times 5 = 20$
7. $4 \times 7 = 28$
8. $3 \times 4 = 12$
9. $8 \times 4 = 32$
10. $4 \times 9 = 36$

11. $4 \times 6 = 24$

12. $4 \times 4 = 16$

13. 4, 8, 12, 16, 20, 24, 28, 32, 36, 40

14. $\dfrac{0}{4 \cdot 0}, \dfrac{4}{4 \cdot 1}, \dfrac{8}{4 \cdot 2}, \dfrac{12}{4 \cdot 3}, \dfrac{16}{4 \cdot 4}, \dfrac{20}{4 \cdot 5},$
$\dfrac{24}{4 \cdot 6}, \dfrac{28}{4 \cdot 7}, \dfrac{32}{4 \cdot 8}, \dfrac{36}{4 \cdot 9}, \dfrac{40}{4 \cdot 10}$

15. $4 \times 8 = 32$ quarters

16. $4 \times 6 = 24$

17. $D \times 4 = 12$
    $D = 3$

18. $10 \times 4 = 40$ quarters

Systematic Review 16D

1. $4 \times 4 = 16$
2. $2 \times 6 = 12$
3. $6 \times 3 = 18$
4. $10 \times 3 = 30$
5. $5 \times 9 = 45$
6. $4 \times 6 = 24$
7. $2 \times 7 = 14$
8. $3 \times 3 = 9$
9. $4 \times 9 = 36$
10. $5 \times 3 - 15$
11. $7 \times 6 = 42$
12. $9 \times 3 = 27$
13. done
14. $3 \times \underline{3} = 9$
15. $4 \times 4 = 16$ sq in
16. $4 + 4 + 4 + 4 = 16$ in
17. $4 \times 8 = 32$ sq mi
18. $8 + 4 + 8 + 4 = 24$ mi
19. $4 \times 9 = 36$ quarters
20. $5 + 2 = 7$
    $7 \times 4 = 28$ qt
    You could also solve this by multiplying
    first to change gallons to quarts.

Systematic Review 16E

1. $6 \times 9 = 54$
2. $7 \times 4 = 28$
3. $8 \times 6 = 48$
4. $10 \times 5 = 50$
5. $4 \times 3 = 12$
6. $9 \times 7 = 63$
7. $1 \times 0 = 0$
8. $9 \times 4 = 36$
9. $1 \times 1 = 1$
10. $6 \times 7 = 42$
11. $9 \times 9 = 81$
12. $4 \times 8 = 32$
13. $5 \times \underline{5} = 25$
14. $\underline{4} \times 10 = 40$
15. $6 \times 6 = 36$ sq in
16. $6 + 6 + 6 + 6 = 24$ in
17. $4 \times 7 = 28$ sq mi
18. $4 + 7 + 4 + 7 = 22$ mi
19. $6 + 2 = 8$ horses
    $8 \times 4 = 32$ horseshoes
20. $\$4 \times 4 = \$16$
    $\$16 + \$16 = \$32$

Systematic Review 16F

1. $7 \times 3 = 21$
2. $8 \times 9 = 72$
3. $5 \times 7 = 35$
4. $6 \times 9 = 54$
5. $2 \times 8 = 16$
6. $9 \times 3 = 27$
7. $7 \times 2 = 14$
8. $7 \times 4 = 28$
9. $5 \times 9 = 45$
10. $10 \times 7 = 70$
11. $6 \times 4 = 24$
12. $1 \times 5 = 5$
13. $2 \times \underline{8} = 16$
14. $3 \times \underline{9} = 27$
15. $16 + 16 + 16 + 16 = 64$ in

16. $8+8+8+8 = 32$ in
17. $5 \times 9 = 45$ sq ft
18. $5+9+5+9 = 28$ ft
19. $5 \times 10¢ = 50¢$; 5 dimes
20. $3+3 = 6$
    $6 \times 9 = 54$ buttons

Lesson Practice 17A

1. 7, 14, 21, 28, 35, 42, 49, 56, 63, 70
2. 7, 14, 21, 28, 35, 42, 49, 56, 63, 70
3. done
4. $\begin{array}{r} 20 \\ \times\ 2 \\ \hline 40 \end{array}$
5. $\begin{array}{r} 40 \\ \times\ 3 \\ \hline 120 \end{array}$
6. 7, 14, 21, 28, $35
7. 7, 14, $21
8. 7, 14, 21, 28, 35, 42 miles

Lesson Practice 17B

1. 7, 14, 21, 28, 35, 42, 49, 56, 63, 70
2. 7, 14, 21, 28, 35, 42, 49, 56, 63, 70
3. $\begin{array}{r} 40 \\ \times\ 5 \\ \hline 200 \end{array}$
4. $\begin{array}{r} 30 \\ \times\ 6 \\ \hline 180 \end{array}$
5. $\begin{array}{r} 20 \\ \times\ 3 \\ \hline 60 \end{array}$
6. 7, 14, 21, 28, 35, 42, 49, 56 trees
7. 7, 14 cookies
8. 7, 14, 21, 28 days

Lesson Practice 17C

1. 7, 14, 21, 28, 35, 42, 49, 56, 63, 70
2. 7, 14, 21, 28, 35, 42, 49, 56, 63, 70
3. $\begin{array}{r} 20 \\ \times\ 6 \\ \hline 120 \end{array}$
4. $\begin{array}{r} 50 \\ \times\ 3 \\ \hline 150 \end{array}$
5. $\begin{array}{r} 80 \\ \times\ 2 \\ \hline 160 \end{array}$
6. 7, 14, 21, 28, 35, 42, 49 dogs
7. 7, 14, 21, 28, 35, 42, 49, 56, 63 bananas
8. 7, 14, 21, 28, 35, $42

Systematic Review 17D

1. 7, 14, 21, 28, 35, 42, 49, 56, 63, 70
2. $\frac{4}{7} = \frac{8}{14} = \frac{12}{21} = \frac{16}{28} = \frac{20}{35}$
3. $4 \times 9 = 36$
4. $3 \times 8 = 24$
5. $2 \times 1 = 2$
6. $3 \times 4 = 12$
7. $7 \times 4 = 28$
8. $8 \times 6 = 48$
9. $\begin{array}{r} 40 \\ \times\ 9 \\ \hline 360 \end{array}$
10. $\begin{array}{r} 50 \\ \times\ 2 \\ \hline 100 \end{array}$
11. $3 \times \underline{6} = 18$
12. $8 \times \underline{10} = 80$
13. $1 \times \underline{5} = 5$
14. $7 \times \underline{9} = 63$
15. done
16. $17 = 17$
17. $31 > 13$
18. $7 \times 4 = 28$ quarters
19. $5+3+5+3 = 16$ in
20. $65-36 = 29$ pennies

Systematic Review 17E

1. 6, 12, 18, 24, 30, 36, 42, 48, 54, 60
2. $\frac{6}{7} = \frac{12}{14} = \frac{18}{21} = \frac{24}{28} = \frac{30}{35}$
3. $4 \times 6 = 24$
4. $4 \times 8 = 32$
5. $6 \times 3 = 18$
6. $9 \times 8 = 72$
7. $4 \times 4 = 16$
8. $8 \times 5 = 40$
9. $\begin{array}{r} 60 \\ \times\ 9 \\ \hline 540 \end{array}$
10. $\begin{array}{r} 30 \\ \times\ 7 \\ \hline 210 \end{array}$
11. $6 \times \underline{9} = 54$
12. $5 \times \underline{2} = 10$
13. $6 \times \underline{0} = 0$
14. $2 \times \underline{6} = 12$
15. 9 qt > 8 qt
16. 18 = 18
17. 15 < 18
18. $32 + 32 + 17 = 81$ hours
19. $5 \times 10 = 50$ years
20. $2 \times 9 = 18$ pt
    $2 \times 6 = 12$ pt
    $18 + 12 = 30$ pints

Systematic Review 17F

1. 7, 14, 21, 28, 35, 42, 49, 56, 63, 70
2. $\frac{4}{5} = \frac{8}{10} = \frac{12}{15} = \frac{16}{20} = \frac{20}{25}$
3. $3 \times 4 = 12$
4. $9 \times 9 = 81$
5. $6 \times 8 = 48$
6. $4 \times 3 = 12$
7. $5 \times 9 = 45$
8. $\begin{array}{r} 10 \\ \times\ 7 \\ \hline 70 \end{array}$

9. $\begin{array}{r} 90 \\ \times\ 3 \\ \hline 270 \end{array}$
10. $\begin{array}{r} 60 \\ \times\ 4 \\ \hline 240 \end{array}$
11. $7 \times \underline{2} = 14$
12. $4 \times \underline{9} = 36$
13. $6 \times \underline{7} = 42$
14. $5 \times \underline{3} = 15$
15. 36 > 30
16. 11 < 12
17. 18 ft = 18 ft
18. $5 \times 10¢ = 50¢$
    $3 \times 5¢ = 15¢$
    $50¢ + 15¢ = 65¢$, <u>no</u>
    $75¢ - 65¢ = \underline{10¢}$
19. $8 \times 4 = 32$ lines
20. $9 \times 10 = 90$ years

Lesson Practice 18A

1. $7 \times 9 = 63$
2. $7 \times 7 = 49$
3. $10 \times 7 = 70$
4. $7 \times 6 = 42$
5. $7 \times 4 = 28$
6. $7 \times 8 = 56$
7. $7 \times 5 = 35$
8. $3 \times 7 = 21$
9. $7 \times 8 = 56$
10. $7 \times 7 = 49$
11.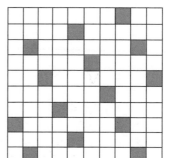

$7 \times 11 = 77$

$7 \times 12 = 84$

$7 \times 13 = 91$

12. done

13.
$$\begin{array}{r} 200 \\ \times\ 4 \\ \hline 800 \end{array}$$

14.
$$\begin{array}{r} 100 \\ \times\ 9 \\ \hline 900 \end{array}$$

15. $7 \times 7 - 49$ skirts

16. $6 \times 7 = 42$ days

Lesson Practice 18B

1. $7 \times 8 = 56$
2. $7 \times 4 = 28$
3. $9 \times 7 = 63$
4. $7 \times 7 = 49$
5. $7 \times 3 = 21$
6. $7 \times 5 = 35$
7. $7 \times 8 = 56$
8. $6 \times 7 = 42$
9.
$$\begin{array}{r} 400 \\ \times\ 2 \\ \hline 800 \end{array}$$
10.
$$\begin{array}{r} 200 \\ \times\ 3 \\ \hline 600 \end{array}$$
11.
$$\begin{array}{r} 100 \\ \times\ 6 \\ \hline 600 \end{array}$$
12.
$$\begin{array}{r} 300 \\ \times\ 2 \\ \hline 600 \end{array}$$
13. 7, 14, 21, 28, 35, 42, 49, 56, 63, 70
14. $\dfrac{0}{7 \times 0}, \dfrac{7}{7 \times 1}, \dfrac{14}{7 \times 2}, \dfrac{21}{7 \times 3}, \dfrac{28}{7 \times 4}, \dfrac{35}{7 \times 5},$

$\dfrac{42}{7 \times 6}, \dfrac{49}{7 \times 7}, \dfrac{56}{7 \times 8}, \dfrac{63}{7 \times 9}, \dfrac{70}{7 \times 10}$
15. $7 \times 7 = 49$
16. $7 \times 8 = 56$ days
17. $7 \times 2 = 14$ days
18. $7 \times 100 = 700$ soldiers

Lesson Practice 18C

1. $7 \times 7 = 49$
2. $7 \times 2 = 14$
3. $7 \times 8 = 56$
4. $9 \times 7 = 63$
5. $5 \times 7 = 35$
6. $6 \times 7 = 42$
7. $7 \times 3 = 21$
8. $7 \times 4 = 28$
9.
$$\begin{array}{r} 200 \\ \times\ 2 \\ \hline 400 \end{array}$$
10.
$$\begin{array}{r} 100 \\ \times\ 5 \\ \hline 500 \end{array}$$
11.
$$\begin{array}{r} 300 \\ \times\ 3 \\ \hline 900 \end{array}$$
12.
$$\begin{array}{r} 200 \\ \times\ 4 \\ \hline 800 \end{array}$$
13. 7, 14, 21, 28, 35, 42, 49, 56, 63, 70
14. $\dfrac{0}{7 \cdot 0}, \dfrac{7}{7 \cdot 1}, \dfrac{14}{7 \cdot 2}, \dfrac{21}{7 \cdot 3}, \dfrac{28}{7 \cdot 4}, \dfrac{35}{7 \cdot 5},$

$\dfrac{42}{7 \cdot 6}, \dfrac{49}{7 \cdot 7}, \dfrac{56}{7 \cdot 8}, \dfrac{63}{7 \cdot 9}, \dfrac{70}{7 \cdot 10}$
15. $7 \times 1 = 7$
16. $4 \times 7 = 28$ slices
17. $7 \times 10 = 70$ times
18. $400 \times 2 = 800$ wings

Systematic Review18D

1. $6 \times 2 = 12$
2. $7 \times 8 = 56$
3. $3 \times 3 = 9$
4. $7 \times 2 = 14$
5. $1 \times 9 = 9$
6. $7 \times 7 = 49$
7. $10 \times 8 = 80$
8. $4 \times 3 = 12$
9.
$$\begin{array}{r} 70 \\ \times\ 6 \\ \hline 420 \end{array}$$

10. $9 \times 7 = 63$

11.
$$\begin{array}{r} 40 \\ \times\ 7 \\ \hline 280 \end{array}$$

12.
$$\begin{array}{r} 200 \\ \times\ 3 \\ \hline 600 \end{array}$$

13. $3 \times \underline{9} = 27$

14. $3 \times \underline{5} = 15$

15. $20 < 21$

16. $31 > 29$

17. $35 = 35$

18. $3 \times 100 = 300$ years old

19. $A = 8 \times 7 = 56$ sq ft
$P = 7 + 8 + 7 + 8 = 30$ ft

20. $79 + 82 + 113 = 274$ cars

Systematic Review18E

1. $9 \times 7 = 63$

2. $2 \times 5 = 10$

3. $3 \times 6 = 18$

4. $7 \times 7 = 49$

5. $2 \times 9 = 18$

6. $8 \times 7 = 56$

7. $5 \times 6 = 30$

8. $10 \times 4 = 40$

9. $10 \times 7 = 70$

10. $6 \times 7 = 42$

11.
$$\begin{array}{r} 60 \\ \times\ 6 \\ \hline 360 \end{array}$$

12.
$$\begin{array}{r} 100 \\ \times\ 8 \\ \hline 800 \end{array}$$

13. $2 \times \underline{7} = 14$

14. $7 \times \underline{7} = 49$

15. $28 = 28$

16. $21 < 22$

17. $25¢ < 30¢$

18. $1 \times 100 = 100$ years old

19. $5 \times 2 = 10$
$15 - 10 = 5$ miles

20. $5 \times 5 = 25$ sq ft
$25 - 11 = 14$ worms

Systematic Review18F

1. $3 \times 7 = 21$

2. $10 \times 7 = 70$

3. $3 \times 4 = 12$

4. $8 \times 7 = 56$

5. $9 \times 9 = 81$

6. $8 \times 4 = 32$

7. $7 \times 7 = 49$

8. $10 \times 2 = 20$

9. $8 \times 9 = 72$

10. $4 \times 7 = 28$

11.
$$\begin{array}{r} 70 \\ \times\ 5 \\ \hline 350 \end{array}$$

12.
$$\begin{array}{r} 200 \\ \times\ 3 \\ \hline 600 \end{array}$$

13. $4 \times \underline{9} = 36$

14. $4 \times \underline{6} = 24$

15. $42 > 40$

16. $360 = 360$

17. $16 = 16$

18. $9 \times 7 = 63$ temperatures

19. $3 + 5 = 8$
$8 \times \$3 = \$24$

20. $3 + 5 = 8$
$8 \times 5 = 40$ miles

Lesson Practice 19A

1. 8, 16, 24, 32, 40, 48, 56, 64, 72, 80

2. see #1

3. $\dfrac{3}{8} = \dfrac{6}{16} = \dfrac{9}{24} = \dfrac{12}{32} = \dfrac{15}{40} =$
$\dfrac{18}{48} = \dfrac{21}{56} = \dfrac{24}{64} = \dfrac{27}{72} = \dfrac{30}{80}$

4. 8, 16, 24, 32, <u>40</u> sides

5. 8, 16, 24, 32, 40, 48, <u>56</u> ; 7 windows

6. 8, 16, 24, 32, 40, <u>48</u> yards

7. 8, 16, 24, 32, 40, 48, 56, <u>64</u> pints

8. 8, 16, 24, <u>32</u> legs

Lesson Practice 19B

1. 8, 16, 24, 32, 40, 48, 56, 64, 72, 80

2. see #1

3. $\frac{6}{8} = \frac{12}{16} = \frac{18}{24} = \frac{24}{32} = \frac{30}{40} =$

   $\frac{36}{48} = \frac{42}{56} = \frac{48}{64} = \frac{54}{72} = \frac{60}{80}$

4. 8, 16, 24, 32, 40, <u>48</u> pints

5. 8, 16, <u>24</u> sides

6. 8, 16, 24, 32, 40, 48, 56, <u>$64</u>

7. 8, <u>16</u> jugs

8. 8, 16, 24, 32, 40, 48, 56, 64, <u>72</u> legs

Lesson Practice 19C

1. 8, 16, 24, 32, 40, 48, 56, 64, 72, 80

2. see #1

3. $\frac{4}{8} = \frac{8}{16} = \frac{12}{24} = \frac{16}{32} = \frac{20}{40} =$

   $\frac{24}{48} = \frac{28}{56} = \frac{32}{64} = \frac{36}{72} = \frac{40}{80}$

4. 8, 16, <u>24</u> legs

5. 8, 16, 24, 32, <u>40</u> tentacles

6. $5+4=9$

   8, 16, 24, 32, 40, 48,
   56, 64, <u>72</u> tentacles

7. 8, 16, 24, 32, 40, 48, 56, 64,
   72, 80; 10 weeks

8. 8, 16, 24, 32, 40, 48, <u>56</u> sides

Systematic Review19D

1. 8, 16, 24, 32, 40, 48, 56, 64, 72, 80

2. $\frac{3}{5} = \frac{6}{10} = \frac{9}{15} = \frac{12}{20} = \frac{15}{25}$

3. $8 \times 3 = 24$

4. $5 \times 3 = 15$

5. $7 \times 7 = 49$

6. $2 \times 4 = 8$

7. $7 \times 8 = 56$

8. $3 \times 6 = 18$

9. $\begin{array}{r} 40 \\ \times\ 8 \\ \hline 320 \end{array}$

10. $\begin{array}{r} 200 \\ \times\ 2 \\ \hline 400 \end{array}$

11. $9 \times \underline{6} = 54$

12. $4 \times \underline{4} = 16$

13. $8 \times \underline{0} = 0$

14. $7 \times \underline{5} = 35$

15. done

16. $\begin{array}{r} 1\,^{1}28 \\ +6\ 35 \\ \hline 7\ 63 \end{array}$

17. $\begin{array}{r} 212 \\ +872 \\ \hline 1,084 \end{array}$

18. 4 couples = 8 people

    8, 16, 24, 32, 40, <u>48</u> dancers

19. $3 \times 3 = 9$ shoes

20. $179 + 143 = 322$ laps

Systematic Review19E

1. 7, 14, 21, 28, 35, 42, 49, 56, 63, 70

2. $\frac{2}{6} = \frac{4}{12} = \frac{6}{18} = \frac{8}{24} = \frac{10}{30}$

3. $8 \times 2 = 16$

4. $5 \times 8 = 40$

5. $9 \times 8 = 72$

6. $3 \times 6 = 18$

7. $7 \times 7 = 49$

8. $10 \times 5 = 50$

9.
$$\begin{array}{r} 60 \\ \times\ 7 \\ \hline 420 \end{array}$$

10.
$$\begin{array}{r} 100 \\ \times\ 9 \\ \hline 900 \end{array}$$

11. $3 \times \underline{9} = 27$

12. $6 \times \underline{6} = 36$

13. $5 \times \underline{1} = 5$

14. $4 \times \underline{5} = 20$

15.
$$\begin{array}{r} 60\ 1 \\ +5\ 13 \\ \hline 1,1\ 14 \end{array}$$

16.
$$\begin{array}{r} {}^1 2\ {}^1 4\ 5 \\ +\ 1\ 89 \\ \hline 4\ 34 \end{array}$$

17.
$$\begin{array}{r} 538 \\ +25\ 1 \\ \hline 789 \end{array}$$

18. $3 + 4 = 7$ gallons

8, 16, 24, 32, 40, 48, $\underline{56}$ pints

19. $2 \times 4 = 8$

$8 + 2 = 10$ lb

20. $74 + 98 + 206 = 378$ coins

Systematic Review19F

1. 8, 16, 24, 32, 40, 48, 56, 64, 72, 80

2. $\dfrac{1}{7} = \dfrac{2}{14} = \dfrac{3}{21} = \dfrac{4}{28} = \dfrac{5}{35}$

3. $8 \times 10 = 80$

4. $6 \times 8 = 48$

5. $8 \times 4 = 32$

6. $3 \times 10 = 30$

7. $1 \times 1 = 1$

8.
$$\begin{array}{r} 80 \\ \times\ 4 \\ \hline 320 \end{array}$$

9.
$$\begin{array}{r} 10 \\ \times\ 5 \\ \hline 50 \end{array}$$

10.
$$\begin{array}{r} 100 \\ \times\ 2 \\ \hline 200 \end{array}$$

11. $8 \times \underline{3} = 24$

12. $7 \times \underline{7} = 49$

13. $4 \times \underline{3} = 12$

14. $7 \times \underline{2} = 14$

15.
$$\begin{array}{r} 4\ {}^1 5\ 2 \\ +3\ 18 \\ \hline 7\ 70 \end{array}$$

16.
$$\begin{array}{r} 7\ 1\ 1 \\ +206 \\ \hline 9\ 17 \end{array}$$

17.
$$\begin{array}{r} {}^1 1\ 5\ 3 \\ +592 \\ \hline 7\ 45 \end{array}$$

18. $\$45 - \$27 = \$18$

19. $6 \times 5 = 30$

$30 - 7 = 23$ gallons

20. $5 \times 10 = 50$ sq ft

$50 - 35 = 15$ tiles

Lesson Practice 20A

1. $8 \times 9 = 72$

2. $7 \times 8 = 56$

3. $6 \times 8 = 48$

4. $8 \times 10 = 80$

5. $5 \times 8 = 40$

6. $1 \times 8 = 8$

7. $8 \times 3 = 24$

8. $8 \times 2 = 16$

9. $8 \times 8 = 64$

10. $8 \times 9 = 72$

11. $4 \times 8 = 32$

12. $8 \times 7 = 56$

13. $8 \times 0 = 0$

14. $8 \times 8 = 64$

15.

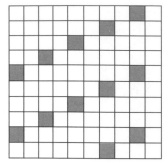

16. $8 \times 10 = 80$ sides
17. $6 \times 8 = 48$ sides
18. $3 \times 8 = 24$ oranges

Lesson Practice 20B

1. $8 \times 4 = 32$
2. $8 \times 8 = 64$
3. $5 \times 8 = 40$
4. $7 \times 8 = 56$
5. $9 \times 8 = 72$
6. $10 \times 8 = 80$
7. $8 \times 6 = 48$
8. $8 \times 3 = 24$
9. $9 \times 8 = 72$
10. $8 \times 2 = 16$
11. $8 \times 8 = 64$
12. $8 \times 1 = 8$
13. 8, 16, 24, 32, 40, 48, 56, 64, 72, 80
14. $\dfrac{0}{(8)(0)}, \dfrac{8}{(8)(1)}, \dfrac{16}{(8)(2)}, \dfrac{24}{(8)(3)},$

$\dfrac{32}{(8)(4)}, \dfrac{40}{(8)(5)}, \dfrac{48}{(8)(6)}, \dfrac{56}{(8)(7)},$

$\dfrac{64}{(8)(8)}, \dfrac{72}{(8)(9)}, \dfrac{80}{(8)(10)}$
15. $5 \times 8 = 40$ sides
16. $8 \times 4 = 32$
17. $8 \times 7 = 56$ slices
18. $9 \times 8 = 72$ people

Lesson Practice 20C

1. $6 \times 8 = 48$
2. $8 \times 2 = 16$
3. $8 \times 8 = 64$
4. $8 \times 9 = 72$
5. $8 \times 3 = 24$
6. $5 \times 8 = 40$
7. $7 \times 8 = 56$
8. $8 \times 10 = 80$
9. $4 \times 8 = 32$
10. $8 \times 1 = 8$
11. $6 \times 8 = 48$
12. $8 \times 9 = 72$
13. $8 \times 8 = 64$
14. $8 \times 5 = 40$
15.

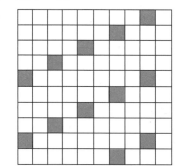

$8 \times 11 = 88$

$8 \times 12 = 96$
16. $8 \times 7 = 56$ sides
17. $8 \times 4 = 32$ sides
18. $8 \times 8 = 64$ cylinders

Systematic Review 20D

1. $7 \times 7 = 49$
2. $8 \times 6 = 48$
3. $9 \times 9 = 81$
4. $3 \times 1 = 3$
5. $8 \times 7 = 56$
6. $9 \times 3 = 27$
7. $7 \times 6 = 42$
8. $9 \times 5 = 45$

9.
$$\begin{array}{r} 80 \\ \times\ 8 \\ \hline 640 \end{array}$$

10. $2 \times 0 = 0$

11.
$$\begin{array}{r} 40 \\ \times\ 4 \\ \hline 160 \end{array}$$

12.
$$\begin{array}{r} 100 \\ \times\ 3 \\ \hline 300 \end{array}$$

13. $49 > 14$

14. $41 = 41$

15. $24 < 28$

16. done

17. done

18.
$$\begin{array}{r} 4\,{}^{1}5\,{}^{5}6\,{}^{1}2 \\ -\ \ 3\ \ 7\ 4 \\ \hline 1\ \ 8\ 8 \end{array}$$

19. $3 + 4 = 7$
$7 \times 8 = 56$ legs

20. $458 - 328 = 130$

Systematic Review 20E

1. $8 \times 8 = 64$
2. $7 \times 7 = 49$
3. $2 \times 6 = 12$
4. $5 \times 4 = 20$
5. $9 \times 4 = 36$
6. $6 \times 6 = 36$
7. $7 \times 8 = 56$
8. $3 \times 7 = 21$
9.
$$\begin{array}{r} 40 \\ \times\ 3 \\ \hline 120 \end{array}$$
10. $2 \times 8 = 16$
11.
$$\begin{array}{r} 80 \\ \times\ 3 \\ \hline 240 \end{array}$$
12.
$$\begin{array}{r} 200 \\ \times\ 2 \\ \hline 400 \end{array}$$
13. $45¢ > 40¢$
14. $24 = 24$

15. $32 < 33$

16.
$$\begin{array}{r} 5\,{}^{9}6\,{}^{9}7\,{}^{1}0\,{}^{1}3 \\ -\ \ 1\ \ \ 1\ 8 \\ \hline 4\ \ \ 8\ 5 \end{array}$$

17.
$$\begin{array}{r} 345 \\ -\ 142 \\ \hline 203 \end{array}$$

18.
$$\begin{array}{r} 8\,{}^{2}3\,{}^{1}7 \\ -\ 1\ 0\ 8 \\ \hline 7\ 2\ 9 \end{array}$$

19. $8 \times 4 = 32$
$8 \times 2 = 16$
$16 + 32 = 48$ tires

20. $244 - 188 = 56$ pencils

Systematic Review 20F

1. $3 \times 6 = 18$
2. $9 \times 7 = 63$
3. $8 \times 4 = 32$
4. $3 \times 9 = 27$
5. $8 \times 8 = 64$
6. $5 \times 0 = 0$
7. $3 \times 3 = 9$
8. $7 \times 7 = 49$
9.
$$\begin{array}{r} 70 \\ \times\ 4 \\ \hline 280 \end{array}$$
10. $9 \times 9 = 81$
11.
$$\begin{array}{r} 60 \\ \times\ 3 \\ \hline 180 \end{array}$$
12.
$$\begin{array}{r} 400 \\ \times\ 2 \\ \hline 800 \end{array}$$
13. $0 < 9$
14. $47 > 18$
15. 16 qt < 20 qt
16.
$$\begin{array}{r} 8\,{}^{9}9\,{}^{9}1\,{}^{1}0\,{}^{1}0 \\ -\ \ 1\ \ \ 2\ 3 \\ \hline 7\ \ \ 7\ 7 \end{array}$$

17.
$$
\begin{array}{r}
6\ ^7\!8\ ^13 \\
-2\ \ 5\ 4 \\
\hline
4\ \ 2\ 9
\end{array}
$$

18.
$$
\begin{array}{r}
^4\!5\ ^1\!06 \\
-\ \ 3\ 44 \\
\hline
1\ 6\ 2
\end{array}
$$

19. $549 + 86 = 635$

$635 - 350 = \$285$

20. $2 \times \underline{9} = 18$; 9 geese

Lesson Practice 21A

1. done

2.
$$
\begin{array}{r}
13 \\
\times\ 2 \\
\hline
26
\end{array}
\qquad
\begin{array}{r}
10+3 \\
\times\ \ \ \ 2 \\
\hline
20+6
\end{array}
$$

3.
$$
\begin{array}{r}
11 \\
\times\ 7 \\
\hline
77
\end{array}
\qquad
\begin{array}{r}
10+1 \\
\times\ \ \ \ 7 \\
\hline
70+7
\end{array}
$$

4.
$$
\begin{array}{r}
21 \\
\times\ 2 \\
\hline
42
\end{array}
\qquad
\begin{array}{r}
20+1 \\
\times\ \ \ \ 2 \\
\hline
40+2
\end{array}
$$

5.
$$
\begin{array}{r}
32 \\
\times\ 3 \\
\hline
96
\end{array}
\qquad
\begin{array}{r}
30+2 \\
\times\ \ \ \ 3 \\
\hline
90+6
\end{array}
$$

6.
$$
\begin{array}{r}
14 \\
\times\ 2 \\
\hline
28
\end{array}
\qquad
\begin{array}{r}
10+4 \\
\times\ \ \ \ 2 \\
\hline
20+8
\end{array}
$$

7.
$$
\begin{array}{r}
11 \\
\times\ 9 \\
\hline
99
\end{array}
\qquad
\begin{array}{r}
10+1 \\
\times\ \ \ \ 9 \\
\hline
90+9
\end{array}
$$

8.
$$
\begin{array}{r}
24 \\
\times\ 2 \\
\hline
48
\end{array}
\qquad
\begin{array}{r}
20+4 \\
\times\ \ \ \ 2 \\
\hline
40+8
\end{array}
$$

9.
$$
\begin{array}{r}
123 \\
\times\ 2 \\
\hline
246
\end{array}
\qquad
\begin{array}{r}
100+20+3 \\
\times\ \ \ \ \ \ \ \ \ \ 2 \\
\hline
200+40+6
\end{array}
$$

10.
$$
\begin{array}{r}
222 \\
\times\ 4 \\
\hline
888
\end{array}
\qquad
\begin{array}{r}
200+20+2 \\
\times\ \ \ \ \ \ \ \ \ \ 4 \\
\hline
800+80+8
\end{array}
$$

11. $4 \times 12 = 48$ $(40+8)$ books

12. $310 \times 3 = 930$ $(900+30)$ shots

Lesson Practice 21B

1.
$$
\begin{array}{r}
21 \\
\times\ 3 \\
\hline
63
\end{array}
\qquad
\begin{array}{r}
20+1 \\
\times\ \ \ \ 3 \\
\hline
60+3
\end{array}
$$

2.
$$
\begin{array}{r}
24 \\
\times\ 2 \\
\hline
48
\end{array}
\qquad
\begin{array}{r}
20+4 \\
\times\ \ \ \ 2 \\
\hline
40+8
\end{array}
$$

3.
$$
\begin{array}{r}
22 \\
\times\ 4 \\
\hline
88
\end{array}
\qquad
\begin{array}{r}
20+2 \\
\times\ \ \ \ 4 \\
\hline
80+8
\end{array}
$$

4.
$$
\begin{array}{r}
11 \\
\times\ 6 \\
\hline
66
\end{array}
\qquad
\begin{array}{r}
10+1 \\
\times\ \ \ \ 6 \\
\hline
60+6
\end{array}
$$

5.
$$
\begin{array}{r}
14 \\
\times\ 2 \\
\hline
28
\end{array}
\qquad
\begin{array}{r}
10+4 \\
\times\ \ \ \ 2 \\
\hline
20+8
\end{array}
$$

6.
$$
\begin{array}{r}
33 \\
\times\ 3 \\
\hline
99
\end{array}
\qquad
\begin{array}{r}
30+3 \\
\times\ \ \ \ 3 \\
\hline
90+9
\end{array}
$$

7.
$$
\begin{array}{r}
110 \\
\times\ 5 \\
\hline
550
\end{array}
\qquad
\begin{array}{r}
100+10+0 \\
\times\ \ \ \ \ \ \ \ \ \ 5 \\
\hline
500+50+0
\end{array}
$$

8.
$$
\begin{array}{r}
231 \\
\times\ 3 \\
\hline
693
\end{array}
\qquad
\begin{array}{r}
200+30+1 \\
\times\ \ \ \ \ \ \ \ \ \ 3 \\
\hline
600+90+3
\end{array}
$$

9.
$$
\begin{array}{r}
424 \\
\times\ 2 \\
\hline
848
\end{array}
\qquad
\begin{array}{r}
400+20+4 \\
\times\ \ \ \ \ \ \ \ \ \ 2 \\
\hline
800+40+8
\end{array}
$$

10.
$$
\begin{array}{r}
121 \\
\times\ 4 \\
\hline
484
\end{array}
\qquad
\begin{array}{r}
100+20+1 \\
\times\ \ \ \ \ \ \ \ \ \ 4 \\
\hline
400+80+4
\end{array}
$$

11. $2 \times 12 = 24$ sq in

12. $2 \times 214 = 428$ legs

Lesson Practice 21C

1.
$$
\begin{array}{r}
43 \\
\times\ 2 \\
\hline
86
\end{array}
\qquad
\begin{array}{r}
40+3 \\
\times\ \ \ \ 2 \\
\hline
80+6
\end{array}
$$

2.
$$
\begin{array}{r}
32 \\
\times\ 2 \\
\hline
64
\end{array}
\qquad
\begin{array}{r}
30+2 \\
\times\ \ \ \ 2 \\
\hline
60+4
\end{array}
$$

3.
$$
\begin{array}{r}
12 \\
\times\ 3 \\
\hline
36
\end{array}
\qquad
\begin{array}{r}
10+2 \\
\times\ \ \ \ 3 \\
\hline
30+6
\end{array}
$$

4.
$$\begin{array}{r} 11 \\ \times\ 4 \\ \hline 44 \end{array} \qquad \begin{array}{r} 10+1 \\ \times\ \ \ \ \ 4 \\ \hline 40+4 \end{array}$$

5.
$$\begin{array}{r} 42 \\ \times\ 2 \\ \hline 84 \end{array} \qquad \begin{array}{r} 40+2 \\ \times\ \ \ \ \ 2 \\ \hline 80+4 \end{array}$$

6.
$$\begin{array}{r} 31 \\ \times\ 3 \\ \hline 93 \end{array} \qquad \begin{array}{r} 30+1 \\ \times\ \ \ \ \ 3 \\ \hline 90+3 \end{array}$$

7.
$$\begin{array}{r} 413 \\ \times\ 2 \\ \hline 826 \end{array} \qquad \begin{array}{r} 400+10+3 \\ \times\ \ \ \ \ \ \ \ \ \ \ 2 \\ \hline 800+20+6 \end{array}$$

8.
$$\begin{array}{r} 111 \\ \times\ 6 \\ \hline 666 \end{array} \qquad \begin{array}{r} 100+10+1 \\ \times\ \ \ \ \ \ \ \ \ \ \ 6 \\ \hline 600+60+6 \end{array}$$

9.
$$\begin{array}{r} 103 \\ \times\ 3 \\ \hline 309 \end{array} \qquad \begin{array}{r} 100+0+3 \\ \times\ \ \ \ \ \ \ \ \ \ 3 \\ \hline 300+0+9 \end{array}$$

10.
$$\begin{array}{r} 212 \\ \times\ 4 \\ \hline 848 \end{array} \qquad \begin{array}{r} 200+10+2 \\ \times\ \ \ \ \ \ \ \ \ \ \ 4 \\ \hline 800+40+8 \end{array}$$

11. $4\times21=84$ plants

12. $8\times111=888$ legs

13.
$$\begin{array}{r} {}^1 23 \\ 45 \\ +\ 17 \\ \hline 85 \end{array}$$

14.
$$\begin{array}{r} {}^2 39 \\ 24 \\ +\ 88 \\ \hline 1\ 51 \end{array}$$

15.
$$\begin{array}{r} 4\,{}^4\cancel{5}\,{}^1 2 \\ -\ 1\ \ 2\ 9 \\ \hline 3\ \ 2\ 3 \end{array}$$

16.
$$\begin{array}{r} 2\,{}^7\cancel{8}\,{}^1 3 \\ -\ 2\ \ 1\ 6 \\ \hline 6\ 7 \end{array}$$

17. $p = 8+11+8+11 = 38$ ft
$a = 8\times11 = 88$ sq ft

18. $9\times7 = 63$
$63+12 = 75$ discoveries

19. $18+52 = 70$ verses

20. $150-123 = 27$ logs

Systematic Review 21D

1.
$$\begin{array}{r} 11 \\ \times\ 5 \\ \hline 55 \end{array} \qquad \begin{array}{r} 10+1 \\ \times\ \ \ \ \ 5 \\ \hline 50+5 \end{array}$$

2.
$$\begin{array}{r} 12 \\ \times\ 3 \\ \hline 36 \end{array} \qquad \begin{array}{r} 10+2 \\ \times\ \ \ \ \ 3 \\ \hline 30+6 \end{array}$$

3.
$$\begin{array}{r} 324 \\ \times\ 2 \\ \hline 648 \end{array} \qquad \begin{array}{r} 300+20+4 \\ \times\ \ \ \ \ \ \ \ \ \ \ 2 \\ \hline 600+40+8 \end{array}$$

4.
$$\begin{array}{r} 322 \\ \times\ 3 \\ \hline 966 \end{array} \qquad \begin{array}{r} 300+20+2 \\ \times\ \ \ \ \ \ \ \ \ \ \ 3 \\ \hline 900+60+6 \end{array}$$

5. $8\times7 = 56$

6. $8\times8 = 64$

7. $7\times7 = 49$

8. $4\times6 = 24$

9. done

10. +

11. −

12. ×

Systematic Review 21E

1.
$$\begin{array}{r} 12 \\ \times\ 4 \\ \hline 48 \end{array} \qquad \begin{array}{r} 10+2 \\ \times\ \ \ \ \ 4 \\ \hline 40+8 \end{array}$$

2.
$$\begin{array}{r} 32 \\ \times\ 3 \\ \hline 96 \end{array} \qquad \begin{array}{r} 30+2 \\ \times\ \ \ \ \ 3 \\ \hline 90+6 \end{array}$$

3.
$$\begin{array}{r} 221 \\ \times\ 4 \\ \hline 884 \end{array} \qquad \begin{array}{r} 200+20+1 \\ \times\ \ \ \ \ \ \ \ \ \ \ 4 \\ \hline 800+80+4 \end{array}$$

4.
$$\begin{array}{r} 313 \\ \times\ 2 \\ \hline 626 \end{array} \qquad \begin{array}{r} 300+10+3 \\ \times\ \ \ \ \ \ \ \ \ \ \ 2 \\ \hline 600+20+6 \end{array}$$

5. $8\times4 = 32$

6. $6\times8 = 48$

7. $7\times9 = 63$

8. $9\times4 = 36$

9. +

10. ×

11. ×

12. −

13.
$$\begin{array}{r} 91 \\ 25 \\ +42 \\ \hline 158 \end{array}$$

14.
$$\begin{array}{r} {}^{1}67 \\ 13 \\ +50 \\ \hline 130 \end{array}$$

15.
$$\begin{array}{r} 8\,{}^{8}9\,{}^{1}3 \\ -6\ \ 15 \\ \hline 2\ \ 78 \end{array}$$

16.
$$\begin{array}{r} {}^{2}\!\cancel{3}\,{}^{1}4 \\ -\ \ \ 92 \\ \hline 222 \end{array}$$

17. $p = 6+7+6+7 = 26$ ft
$a = 6 \times 7 = 42$ sq ft

18. $200 \times 3 = 600$ fish

19. $3 \times 12 = 36$ doughnuts
$36 + 36 = 72$ doughnuts

20. $17 + 22 + 24 + 31 = 94$ pushups

13.
$$\begin{array}{r} {}^{1}82 \\ 13 \\ +56 \\ \hline 151 \end{array}$$

14.
$$\begin{array}{r} {}^{1}71 \\ 64 \\ +26 \\ \hline 161 \end{array}$$

15.
$$\begin{array}{r} 5\,{}^{5}\cancel{6}\,{}^{1}\!\cancel{1} \\ -\ \ \ 19 \\ \hline 5\ \ 42 \end{array}$$

16.
$$\begin{array}{r} {}^{3}\!\cancel{4}\,{}^{1}37 \\ -\ \ 365 \\ \hline 72 \end{array}$$

17. $15 + 12 + 15 + 12 = 54$ ft

18. $2 \times 431 = 862$ birds

19. $7 + 2 = 9$ fish per hour
$9 \times 6 = 54$ fish

20. $7 \times 11 = 77$ eggs
$77 - 18 = 59$ left over

Systematic Review 21F

1.
$$\begin{array}{r} 21 \\ \times\ 2 \\ \hline 42 \end{array} \qquad \begin{array}{r} 20+1 \\ \times\ \ \ \ 2 \\ \hline 40+2 \end{array}$$

2.
$$\begin{array}{r} 11 \\ \times\ 6 \\ \hline 66 \end{array} \qquad \begin{array}{r} 10+1 \\ \times\ \ \ \ 6 \\ \hline 60+6 \end{array}$$

3.
$$\begin{array}{r} 202 \\ \times\ 3 \\ \hline 606 \end{array} \qquad \begin{array}{r} 200+0+2 \\ \times\ \ \ \ \ \ \ \ 3 \\ \hline 600+0+6 \end{array}$$

4.
$$\begin{array}{r} 444 \\ \times\ 2 \\ \hline 888 \end{array} \qquad \begin{array}{r} 400+40+4 \\ \times\ \ \ \ \ \ \ \ \ \ 2 \\ \hline 800+80+8 \end{array}$$

5. $8 \times 6 = 48$

6. $3 \times 7 = 21$

7. $6 \times 10 = 60$

8. $9 \times 7 = 63$

9. −

10. ×

11. +

12. ×

Lesson Practice 22A

1. 40

2. 10

3. 60

4. 60

5. 40

6. 60

7. 400

8. 200

9. 700

10. 2,000

11. 2,000

12. 7,000

13. done

14. $30 \times 6 = 180$

15. $20 \times 3 = 60$

16. done

17. $400 \times 3 = 1,200$

18. $100 \times 4 = 400$

19. $30 \times 5 = 150$ cars

20. $40 \times 3 = \$120$

Lesson Practice 22B

1. 30
2. 70
3. 50
4. 500
5. 700
6. 200
7. 900
8. 100
9. 200
10. 6,000
11. 9,000
12. 4,000
13. $10 \times 2 = 20$
14. $30 \times 2 = 60$
15. $20 \times 7 = 140$ ~~140~~
16. $400 \times 3 = 1,200$
17. $100 \times 4 = 400$
18. $600 \times 2 = 1,200$
19. $20 \times 5 = 100$ mi
20. $50 \times 3 = \$150$

Lesson Practice 22C

1. 90
2. 40
3. 20
4. 200
5. 500
6. 900
7. 400
8. 200
9. 600
10. 8,000
11. 1,000
12. 3,000
13. $50 \times 2 = 100$
14. $40 \times 3 = 120$
15. $20 \times 5 = 100$
16. $100 \times 8 = 800$
17. $300 \times 2 = 600$

18. $400 \times 6 = 2,400$
19. $70 \times 9 = 630$ books
20. $20 \times 4 = 80$ people

Systematic Review 22D

1. 70
2. 10
3. 50
4. $10 \times 8 = 80$
5. $80 \times 9 = 720$
6. $60 \times 5 = 300$
7. $500 \times 3 = 1,500$
8. $400 \times 6 = 2,400$
9. $900 \times 2 = 1,800$
10.

$$\begin{array}{r} 11 \\ \times\ 6 \\ \hline 66 \end{array} \qquad \begin{array}{r} 10+1 \\ \times\ \ \ 6 \\ \hline 60+6 \end{array}$$

11.

$$\begin{array}{r} 21 \\ \times\ 4 \\ \hline 84 \end{array} \qquad \begin{array}{r} 20+1 \\ \times\ \ \ 4 \\ \hline 80+4 \end{array}$$

12.

$$\begin{array}{r} 113 \\ \times\ \ 3 \\ \hline 339 \end{array} \qquad \begin{array}{r} 100+10+3 \\ \times\ \ \ \ \ \ \ 3 \\ \hline 300+30+9 \end{array}$$

13.

$$\begin{array}{r} 423 \\ \times\ \ 2 \\ \hline 846 \end{array} \qquad \begin{array}{r} 400+20+3 \\ \times\ \ \ \ \ \ \ 2 \\ \hline 800+40+6 \end{array}$$

14. $56 > 54$
15. 6 qt < 8 qt
16. $28 > 24$
17. $200 \times 3 = \$600$
18. $53 \times 2 = 106$ jars
19. $4 \times 12 = 48$ soldiers
20. $2 \times 7 = 14$; $3 \times 3 = 9$
    $14 + 9 = 23$ pts Falcons
    $1 \times 7 = 7$; $6 \times 3 = 18$
    $7 + 18 = 25$ pts 49ers
    $23 < 25$, so 49ers won
    Cammi

Systematic Review 22E

1. 400
2. 200
3. 600
4. $20 \times 4 = 80$
5. $70 \times 5 = 350$
6. $90 \times 6 = 540$
7. $200 \times 8 = 1,600$
8. $500 \times 4 = 2,000$
9. $400 \times 3 = 1,200$

10.
$$
\begin{array}{rr}
44 & 40+4 \\
\times\ 2 & \times\ \ \ \ 2 \\
\hline
88 & 80+8
\end{array}
$$

11.
$$
\begin{array}{rr}
32 & 30+2 \\
\times\ 2 & \times\ \ \ \ 2 \\
\hline
64 & 60+4
\end{array}
$$

12.
$$
\begin{array}{rr}
303 & 300+0+3 \\
\times\ 3 & \times\ \ \ \ \ \ \ \ 3 \\
\hline
909 & 900+0+9
\end{array}
$$

13.
$$
\begin{array}{rr}
122 & 100+20+2 \\
\times\ 4 & \times\ \ \ \ \ \ \ \ \ \ 4 \\
\hline
488 & 400+80+8
\end{array}
$$

14. $64 > 60$
15. $40¢ = 40¢$
16. $40 \text{ pt} < 45 \text{ pt}$
17. $2 \times 12 = 24$ pints
18. $2 \times 23 = 46$
    $46 + 24 = 70$ pints
19. $13 + 12 + 13 + 12 = 50$
    $50 = 50$; yes
20. $\$50 - \$19 = \$31$

Systematic Review 22F

1. 3,000
2. 4,000
3. 8,000
4. $80 \times 7 = 560$
5. $60 \times 3 = 180$
6. $50 \times 4 = 200$
7. $900 \times 3 = 2,700$
8. $700 \times 2 = 1,400$
9. $300 \times 5 = 1,500$

10.
$$
\begin{array}{rr}
13 & 10+3 \\
\times\ 3 & \times\ \ \ \ 3 \\
\hline
39 & 30+9
\end{array}
$$

11.
$$
\begin{array}{rr}
41 & 40+1 \\
\times\ 2 & \times\ \ \ \ 2 \\
\hline
82 & 80+2
\end{array}
$$

12.
$$
\begin{array}{rr}
111 & 100+10+1 \\
\times\ 8 & \times\ \ \ \ \ \ \ \ \ 8 \\
\hline
888 & 800+80+8
\end{array}
$$

13.
$$
\begin{array}{rr}
323 & 300+20+3 \\
\times\ 3 & \times\ \ \ \ \ \ \ \ \ \ 3 \\
\hline
969 & 900+60+9
\end{array}
$$

14. $24 = 24$
15. $56¢ < 60¢$
16. $36 > 32$
17. $24 + 4 + 3 + 1 = 32$ animals
18. $\$2 \times 24 = \$48$
    $\$12 \times 4 = \$48$
    $\$6 \times 3 = \$18$
    $\$28 \times 1 = \$28$
    $48 + 48 + 18 + 28 = \$142$
19. $70 \times 6 = 420$ gallons
20. $50 \times 7 = 350$ gallons
    $420 > 350$

Lesson Practice 23A

1. done
2. $11 \times 11 = 121$
3. done

4.
$$
\begin{array}{rr}
32 & 30+2 \\
\times 11 & \times 10+1 \\
\hline
32 & 30+2 \\
32\ \ & 300+20\ \ \ \\
\hline
352 & 300+50+2
\end{array}
$$

5.
$$
\begin{array}{rr}
22 & 20+2 \\
\times 10 & \times 10+0 \\
\hline
0 & +0 \\
22\ \ & 200+20\ \ \ \\
\hline
220 & 200+20+0
\end{array}
$$

6.
$$
\begin{array}{rr}
23 & 20+3 \\
\times 13 & \times 10+3 \\
\hline
69 & 60+9 \\
23\ \ & 200+30\ \ \ \\
\hline
299 & 200+90+9
\end{array}
$$

7.
```
    12          10+2
  × 12        × 10+2
    24          20+4
   12         100+20
  144         100+40+4
```

8.
```
    21          20+1
  × 14        × 10+4
    84          80+4
   21         200+10
  294         200+90+4
```

9. $12×13 = $156

10. 17×11 = 187 pets

Lesson Practice 23B

1. 12×22 = 264

2. 12×13 = 156

3.
```
    13          10+3
  × 11        × 10+1
    13          10+3
   13         100+30
  143         100+40+3
```

4.
```
    21          20+1
  × 12        × 10+2
    42          40+2
   21         200+10
  252         200+50+2
```

5.
```
    45          40+5
  × 11        × 10+1
    45          40+5
   45         400+50
  495         400+90+5
```

6.
```
    23          20+3
  × 21        × 20+1
    23          20+3
   46         400+60
  483         400+80+3
```

7.
```
    22          20+2
  × 13        × 10+3
    66          60+6
   22         200+20
  286         200+80+6
```

8.
```
    37          30+7
  × 10        × 10+0
     0          +0
   37         300+70
  370         300+70+0
```

9. 20×12 = 240 months

10. 11×41 = 451 plants

Lesson Practice 23C

1. 21×11 = 231

2. 11×12 = 132

3.
```
    22          20+2
  × 11        × 10+1
    22          20+2
   22         200+20
  242         200+40+2
```

4.
```
    33          30+3
  × 12        × 10+2
    66          60+6
   33         300+30
  396         300+90+6
```

5.
```
    11          10+1
  × 16        × 10+6
    66          60+6
   11         100+10
  176         100+70+6
```

6.
```
    19          10+9
  × 10        × 10+0
     0          +0
   19         100+90
  190         100+90+0
```

7.
```
    22          20+2
  × 22        × 20+2
    44          40+4
   44         400+40
  484         400+80+4
```

8.
```
    44          40+4
  × 11        × 10+1
    44          40+4
   44         400+40
  484         400+80+4
```

9. 15×11 = 165 birds

10. 13×11 = 143 sq ft

Systematic Review 23D

1.
```
    20
  ×13
    60
    20
   260
```

2.
```
    27
  ×11
    27
    27
   297
```

3.
```
    13
  ×13
    39
    13
   169
```

4.
```
    24
  ×21
    24
  ¹48
  5 04
```

5.
```
    12
  ×31
    12
    36
   372
```

6.
```
    23
  ×12
    46
    23
   276
```

7.
```
    12
  ×44
    48
  ¹48
  5 28
```

8.
```
    31
  ×21
    31
    62
   651
```

9. 600

10. 300

11. 800

12. $20 \times 9 = 180$

13. $30 \times 7 = 210$

14. $30 \times 6 = 180$

15.
```
  ⁴5 ¹9
  -  124
     395
```

16.
```
  ⁷8 ¹63
  -  2 93
     5 70
```

17.
```
  ⁵6 ¹⁰¹₁
  -  2  99
     3   12
```

18. $113 \times 2 = 226$ ants

19. $212 \times 4 = 848$ salmon

20. $100 \times 2 = 200$ pints

Systematic Review 23E

1.
```
    30
  ×13
    90
    30
   390
```

2.
```
    21
  ×20
     0
    42
   420
```

3.
```
    31
  ×12
    62
    31
   372
```

4.
```
    44
  ×22
    88
  1
    88
   968
```

5.
```
    22
  ×33
    66
  1
    66
  7 26
```

6.
```
    13
  ×11
    13
   13
  143
```

7.
```
    21
  ×22
    42
   42
  462
```

8.
```
    10
  ×21
    10
   20
  210
```

9. 90

10. 40

11. 30

12. $700 \times 3 = 2{,}100$

13. $100 \times 5 = 500$

14. $300 \times 8 = 2{,}400$

15.
```
   0 1
   1 16
  -  9 5
     2 1
```

16.
```
   6  9 1 0 1
   7  0 0 0
  -  6  0 2
        9 8
```

17.
```
   5 4 5 1
   5  5 8
  -3  4 9
   2  0 9
```

18. $44 + 26 = 70$

   $70 - 21 = 49$ tons

19. $40 + 50 = \$90$

20. $\$60 \times 9 = \$540$

Systematic Review 23F

1.
```
    35
  ×11
    35
   35
  385
```

2.
```
    23
  ×10
     0
   23
  230
```

3.
```
    26
  ×11
    26
   26
  286
```

4.
```
    22
  ×13
    66
   22
  286
```

5.
```
    44
  ×11
    44
   44
  484
```

6.
```
    23
  ×12
    46
   23
  276
```

7.
```
    12
  ×13
    36
   12
  156
```

8.
```
    20
  ×44
    80
   80
  880
```

9. 7,000

10. 1,000

11. 5,000

12. $900 \times 8 = 7{,}200$

13. $600 \times 9 = 5{,}400$

14. $800 \times 7 = 5{,}600$

15.
```
   1 2 1
     2 98
  + 1 63
    4 6 1
```

16.
```
   1
   562
  + 475
  1,037
```

17.
```
   8 ¹09
  +9 17
  1,7 26
```

18. $236+349=585$

    $585-490=95$ plants

19. $12\times24=288$ pencils

20. $230\times3=690$ people

Lesson Practice 24A

1. done

2.
```
   24        20+4
  ×18       ×10+8
   13      100  30
  162      100+60+2
   24      200+40
  432      400+30+2
```

3.
```
   22        20+2
  ×26       ×20+6
    1            10
  122     100  20+2
   44     400+40
  572     500+70+2
```

4.
```
   46        40+6
  ×12       ×10+2
   1¹           10
   82     100  80+2
   46     400+60
  552     500+50+2
```

5.
```
   27        20+7
  ×16       ×10+6
   14      100  40
  122      100  20+2
   27      200+70
  432      400+30+2
```

6.
```
   36        30+6
  ×24       ×20+4
    2            20
  124      100  20+4
    1      100
   62      600+20
  864      800+60+4
```

7.
```
   35        30+5
  ×29       ×20+9
   14      100  40
  275      200+70+5
    1      100
   60      600+ 0
 1,015   1,000+ 0 +10+5
```

8.
```
   25        20+5
  ×23       ×20+3
    1            10
   65       60+5
    1      100
   40      400
  575      500+70+5
```

9.
```
   35        30+5 ;
  ×15       ×10+5
   12      100  20
  155      100+50+5
   35      300+50
  525      500+20+5     525 cars
```

10.
```
   25        20+5 ;
  ×12       ×10+2
   11      100  10
   40       40+0
   25      200+50
  300      300+ 0 +0    300 tons
```

Lesson Practice 24B

1.
```
   19        10+9
  ×32       ×30+2
    1      100  10
  1 28       20+8
    2      200
   37      300+70
  608      600+ 0 +8
```

2.
```
   42        40+2
  ×66       ×60+6
    1            10
  242      200+40+2
    1      100
  242      2000+400+20
 2772      2000+700+70+2
```

3.
```
    33        30+3
   ×26       ×20+6
    11      100  10
   188      100+80+8
    66      600+60
   858      800+50+8
```

4.
```
    37        30+7
   ×12       ×10+2
    11      100  10
    64       60+4
    37      300+70
   444      400+40+4
```

5.
```
    13        10+3
   ×19       ×10+9
    12        20
           100
    97         90+7
    13      100+30
   247      200+40+7
```

6.
```
    16        10+6
   ×29       ×20+9
     5      100  50
   194        90+4
    22      200+20
   464      400+60+4
```

7.
```
    34        30+4
   ×17       ×10+7
     2        20
   218      200+10+8
    34      300+40
   578      500+70+8
```

8.
```
    48           40+8
   ×26          ×20+6
     4           40
   248         200+40+8
     1         100
    86         800+60
  1248      1,000+200+40+8
```

9.
```
    35        30+5; $455
   ×13       ×10+3
     1      100  10
    95         90+5
     1      300+50
    35      400+50+5
   455
```

10.
```
    45        40+5
   ×24       ×20+4
     2        20
   160      100+60+0
     1      100
    80      800+ 0
  1080    1,000+ 0 +80+0
```

1,080 mosquitoes

Lesson Practice 24C

1.
```
    23        20+3
   ×14       ×10+4
     1        10
    82        80+2
    23      100
   322      200+30
           300+20+2
```

2.
```
    27        20+7
   ×16       ×10+6
    14      100  40
   122      100+20+2
    27      200+70
   432      400+30+2
```

3.
```
    29        20+9
   ×22       ×20+2
     1      100  10
   ¹148      100+40+8
    48      400+80
   638      600+30+8
```

4.
```
    35        30+5
   ×15       ×10+5
    12      100  20
   155      100+50+5
    35      300+50
   525      500+20+5
```

5.
```
    28        20+8
   ×22       ×20+2
    11      100  10
    46        40+6
     1      100
    46      400+60
   616      600+10+6
```

6.
```
     36          30+6
   ×24         ×20+4
     2            20
   124       100+20+4
   1            100
    62        600+20
   864       800+60+4
```

7.
```
     44                40+4
   ×27              ×20+7
   1 2            100  20
   288          200+80+8
    88            800+80
  1188    1,000+100+80+8
```

8.
```
     56          50+6
   ×16         ×10+6
     3            30
   306       300+0  +6
    56        500+60
   896       800+90+6
```

9.
```
     75       70+5 ; 975 lb
   ×13         ×10+3
     1            10
   215       200+10+5
    75        700+50
   975       900+70+5
```

10.
```
     63                60+3
   ×25              ×20+5
     1                  10
   305          300+ 0 +5
   126       1,000+200+60
  1575     1,000+500+70+5

  $1,575
```

Systematic Review 24D

1.
```
     3 2
   × 1 7
      1
   2 14
   3 2
   5 44
```

2.
```
    14
   ×28
    3
  ¹82
   28
  392
```

3.
```
    36
  × 2 2
   1 1
    62
   6 2
  7 92
```

4.
```
    41
   ×38
   328
   123
  1558
```

5.
```
    50
   ×32
   100
   150
  1600
```

6.
```
    31
  × 33
    93
  ¹93
  1 023
```

7.
```
    12
   ×41
    12
    48
   492
```

8.
```
    43
  × 12
    86
  ¹43
   516
```

9. $100 \times 4 = 400$
10. $60 \times 8 = 480$
11. $30 \times 7 = 210$
12. 2, 4, 6, 8, 10, 12, 14, 16, 18, 20
13. $3 \times \underline{6} = 18$
14. $3 \times \underline{8} = 24$
15. $8 \times \underline{4} = 32$
16. $15 \times 15 = 225$ sq ft
17. $13 \times 52 = 676$ hours

18. $240+335+124=699$ yd
19. $61+57=118$
    $118-59=59$ clues
20. $2\times5=10$
    $10\times8=80$ plants

Systematic Review 24E

1. $\begin{array}{r} 18 \\ \times\,40 \\ \hline 0 \\ 3 \\ 42 \\ \hline 720 \end{array}$

2. $\begin{array}{r} 39 \\ \times23 \\ \hline 12 \\ 197 \\ 68 \\ \hline 897 \end{array}$

3. $\begin{array}{r} 36 \\ \times99 \\ \hline 15 \\ 274 \\ 5 \\ 274 \\ \hline 3564 \end{array}$

4. $\begin{array}{r} 76 \\ \times89 \\ \hline 15 \\ 634 \\ 14 \\ 568 \\ \hline 6764 \end{array}$

5. $\begin{array}{r} 24 \\ \times21 \\ \hline 24 \\ {}^{1}48 \\ \hline 504 \end{array}$

6. $\begin{array}{r} 21 \\ \times14 \\ \hline 84 \\ 21 \\ \hline 294 \end{array}$

7. $\begin{array}{r} 23 \\ \times22 \\ \hline 46 \\ 1 \\ 46 \\ \hline 506 \end{array}$

8. $\begin{array}{r} 12 \\ \times13 \\ \hline 36 \\ 12 \\ \hline 156 \end{array}$

9. $600\times3=1,800$
10. $100\times5=500$
11. $800\times8=6,400$
12. 3, 6, 9, 12, 15, 18, 21, 24, 27, 30
13. $4\times\underline{9}=36$
14. $5\times\underline{5}=25$
15. $2\times\underline{6}=12$
16. $47\times31=1,457$ sq mi
    $1,457-100=1,357$ sq mi
17. 3 doz $+$ 2 doz $=$ 5 doz
    $5\times12=60$ eggs
18. $11\times25=275$ cents
19. 300 yards $=$ 900 feet
20. $7\times8=56$
    $56\times3=168$ slices

Systematic Review 24F

1. $\begin{array}{r} 41 \\ \times62 \\ \hline 82 \\ 2{}^{1}46 \\ \hline 2542 \end{array}$

2. $\begin{array}{r} 55 \\ \times25 \\ \hline 2 \\ 255 \\ 1 \\ 100 \\ \hline 1375 \end{array}$

3.
```
      53
    ×35
      11
     255
     159
    1855
```

4.
```
      28
    × 38
      16
     164
       2
      64
    1064
```

5.
```
      49
    ×38
      17
     322
       2
     127
    1862
```

6.
```
      12
    ×42
      24
       1
      48
     504
```

7.
```
      24
    ×22
      48
    ¹48
     528
```

8.
```
      55
    ×11
      55
    ¹55
     605
```

9. $200×4=800$

10. $300×9=2,700$

11. $300×3=900$

12. 4, 8, 12, 16, 20, 24, 28, 32, 36, 40

13. $10×\underline{6}=60$

14. $4×\underline{7}=28$

15. $3×\underline{5}=15$

16. $2×12=24$ cookies

17. $2×18=36$
    $5×4=20$
    $20+36=56$ wheels

18. $10×12=120$ books

19. $23×2=46$
    $3×3=9$
    $46+9=55$ points

20. $82+140=222$ gal
    $222×4=888$ qt

Lesson Practice 25A

1. done

2.
```
       (600)          624
    ×   (80)        × 81
    (48,000)          624
                      ⁴3
                    ¹4862
                    50,544
```

3.
```
       (300)          305
    ×   (20)        × 21
    (6,000)           305
                      610
                    6,405
```

4.
```
       (300)          319
    ×   (30)        × 33
     (900)           12₇
                      937
                       2
                   ¹ 937
                   10,527
```

5.
```
       (500)          495
    ×   (70)        × 72
    (35,000)            1
                      111
                      880
                      163
                     2835
                    35,640
```

6.　　　(900)　　　876
　　× 　　(20)　　× 19
　　　(18,000)　　　1
　　　　　　　　　165
　　　　　　　　7234
　　　　　　　　 876
　　　　　　　　16,644

7.　　　(400)　　　352
　　× 　　(30)　　× 25
　　　(12,000)　　　1
　　　　　　　　　21
　　　　　　　　1 550
　　　　　　　　　1
　　　　　　　　 604
　　　　　　　　8,800

8.　　　(700)　　　681
　　× 　　(40)　　× 38
　　　(28,000)　　1 6
　　　　　　　　4848
　　　　　　　　1 2
　　　　　　　　1 8 43
　　　　　　　　2 5,878

9.　　　(500)　　　500 ; 9,000 toothpicks
　　× 　　(20)　　× 18
　　　(10,000)　　4000
　　　　　　　　 500
　　　　　　　　9,000

10.　× 　　(20)　　　× 23
　　　(12,000)　　　1 1
　　　　　　　　　1826
　　　　　　　　1 284
　　　　　　　　1 4,766

11.　　　(200)　　　212 ; 5,300 gallons
　　× 　　(30)　　× 25
　　　(6,000)　　　1 1
　　　　　　　　1 050
　　　　　　　　 424
　　　　　　　　5,300

12.　　　(400)　　　352 ; 5,280 pages
　　× 　　(20)　　× 15
　　　(8,000)　　　1 2 1
　　　　　　　　1 550
　　　　　　　　 352
　　　　　　　　5,280

Lesson Practice 25B

1.　　　(100)　　　125
　　× 　　(30)　　× 25
　　　(3,000)　　　1 2
　　　　　　　　　505
　　　　　　　　1 1
　　　　　　　　2 40
　　　　　　　　3,125

2.　　　(700)　　　681
　　× 　　(40)　　× 38
　　　(28,000)　　1 6
　　　　　　　　4848
　　　　　　　　1 2
　　　　　　　　1843
　　　　　　　　25,878

3.　　　(500)　　　492
　　× 　　(80)　　× 75
　　　(40,000)　　1 1
　　　　　　　　2 450
　　　　　　　　1 6 1
　　　　　　　　2 8 34
　　　　　　　　36,900

4.　　　(100)　　　145
　　× 　　(20)　　× 15
　　　(2,000)　　　2
　　　　　　　　　525
　　　　　　　　1
　　　　　　　　1 45
　　　　　　　　2,175

5.　　　(500)　　　534
　　× 　　(40)　　× 44
　　　(20,000)　　1 1
　　　　　　　　2 026
　　　　　　　　1 1
　　　　　　　　2026
　　　　　　　　23,496

6.　　　(700)　　　719
　　× 　　(100)　　× 99
　　　(70,000)　　218
　　　　　　　　6391
　　　　　　　　1 8
　　　　　　　　6391
　　　　　　　　71,181

7.
```
      (200)        1 5 4
  ×    (20)       ×  16
    (4,000)           3
                   6 2 4
                       1
                   1 5 4
                   2, 464
```

8.
```
      (300)          290
  ×    (40)         ×41
   (12,000)          290
                       3
                     860
                  11,890
```

9.
```
      (600)          582
  ×    (70)         ×73
   (42,000)          1 2
                   15 4 6
                   1 5 1
                   3 5 6 4
                  42,486
```
42,486 new mosquitoes

10.
```
      (200)          235
  ×    (10)         ×13
    (2,000)          1 1 1
                     695
                   2 3 5
                   3,055
```
$3,055

11.
```
      (400)          365
  ×    (40)         ×35
   (16,000)          1 3
                   1 5 2 5
                   1 1
                     9 8 5
                  12,775
```
12,775 pennies

12.
```
      (400)          365
  ×    (70)         ×70
   (28,000)            0
                      43
                   2 1 2 5
                  25,550
```
25,550 pennies

Lesson Practice 25C

1.
```
      (800)          816
  ×    (80)         ×79
   (64,000)        1 1 5  4
                   7 2 9
                   1  4
                   5 6 7 2
                  64,464
```

2.
```
      (500)          492
  ×    (60)         ×55
   (30,000)          1 1
                   2 4 5 0
                       1
                   2 4 5 0
                  27,060
```

3.
```
      (400)          373
  ×    (60)         ×64
   (24,000)            1
                      21
                   1 2 8 2
                   1 4 1
                   1 8 2 8
                  23,872
```

4.
```
      (800)          777
  ×    (30)         ×33
   (24,000)           22
                   2 1 1 1
                   2 2
                   2 1 1 1
                  25,641
```

5.
```
      (400)          436
  ×    (40)         ×36
   (16,000)            1
                   1 1 3
                   2 4 8 6
                       1
                   1 2 9 8
                  15,696
```

6.
```
      (900)          947
  ×    (10)         ×14
    (9,000)           1 2
                   3 6 6 8
                     9 4 7
                  13,258
```

7.
$$
\begin{array}{r}
(600) \\
\times\ (60) \\
\hline
(36{,}000)
\end{array}
\qquad
\begin{array}{r}
559 \\
\times 63 \\
\hline
1 \\
112 \\
1557 \\
3354 \\
\hline
35{,}217
\end{array}
$$

8.
$$
\begin{array}{r}
(500) \\
\times\ (50) \\
\hline
(25{,}000)
\end{array}
\qquad
\begin{array}{r}
519 \\
\times 52 \\
\hline
1 \\
1028 \\
4 \\
2555 \\
\hline
26{,}988
\end{array}
$$

9.
$$
\begin{array}{r}
(1{,}000) \\
\times\ (100) \\
\hline
(100{,}000)
\end{array}
\qquad
\begin{array}{r}
963;\ 94{,}374\ \text{sq ft} \\
\times 98 \\
\hline
142 \\
7284 \\
152 \\
8147 \\
\hline
94{,}374
\end{array}
$$

10.
$$
\begin{array}{r}
(300) \\
\times\ (50) \\
\hline
(15{,}000)
\end{array}
\qquad
\begin{array}{r}
279;\ \$14{,}508 \\
\times 52 \\
\hline
1 \\
11 \\
448 \\
14 \\
1355 \\
\hline
14{,}508
\end{array}
$$

11.
$$
\begin{array}{r}
(100) \\
\times\ (10) \\
\hline
(1{,}000)
\end{array}
\qquad
\begin{array}{r}
105;\ 1{,}260\ \text{pencils} \\
\times 12 \\
\hline
210 \\
105 \\
\hline
1{,}260
\end{array}
$$

12.
$$
\begin{array}{r}
(200) \\
\times\ (10) \\
\hline
(2{,}000)
\end{array}
\qquad
\begin{array}{r}
215;\ 3{,}010\ \text{mi} \\
\times 14 \\
\hline
112 \\
840 \\
215 \\
\hline
3{,}010
\end{array}
$$

Systematic Review 25D

1.
$$
\begin{array}{r}
873 \\
\times 30 \\
\hline
2\ \ 0 \\
2419 \\
\hline
26{,}190
\end{array}
$$

2.
$$
\begin{array}{r}
718 \\
\times 38 \\
\hline
116\,_4 \\
568 \\
2 \\
2134 \\
\hline
27{,}284
\end{array}
$$

3.
$$
\begin{array}{r}
314 \\
\times 35 \\
\hline
2 \\
1550 \\
1 \\
932 \\
\hline
10{,}990
\end{array}
$$

4.
$$
\begin{array}{r}
85 \\
\times 36 \\
\hline
13\,_0 \\
48 \\
11 \\
245 \\
\hline
3{,}060
\end{array}
$$

5.
$$
\begin{array}{r}
42 \\
\times 59 \\
\hline
1 \\
368 \\
210 \\
\hline
2{,}478
\end{array}
$$

6.
$$
\begin{array}{r}
61 \\
\times 47 \\
\hline
427 \\
244 \\
\hline
2{,}867
\end{array}
$$

7. $11 \times 33 = 363$ sq ft
8. $11 + 33 + 11 + 33 = 88$ ft
9. $12 \times 14 = 168$ sq in
10. $12 + 14 + 12 + 14 = 52$ in
11. $25 \times 25 = 625$ sq in
12. $25 \times 4 = 100$ in
13. $42 > 40$
14. $36 = 36$

15. $15 < 16$
16. 5, 10, 15, 20, 25, 30, 35, 40, 45, 50
17. $21 \times 36 = 756$ pieces
18. $10,000 \times 4 = \$40,000$
19. $43 + 18 = 61$

    $61 - 55 = 6$ cars
20. $12 \times 3 = 36$ feet

Systematic Review 25E

1.
```
   522
  × 93
    11
  1566
    11
  4588
 48,546
```

2.
```
   832
  × 57
   121
  5614
    1
  4150
 47,424
```

3.
```
   471
  × 84
    1
   12
  1684
    5
  3268
 39,564
```

4.
```
   347
  ×  8
    35
  2426
  2,776
```

5.
```
    43
  × 25
   215
    86
  1,075
```

6.
```
    24
  × 31
    24
    1
    62
   744
```

7. $21 \times 40 = 840$ sq ft
8. $21 + 40 + 21 + 40 = 122$ ft
9. $7 \times 13 = 91$ sq in
10. $7 + 13 + 7 + 13 = 40$ in
11. $15 \times 15 = 225$ sq in
12. $15 \times 4 = 60$ in
13. $81 > 80$
14. $48 < 50$
15. $66 > 64$
16. 6, 12, 18, 24, 30, 36, 42, 48, 54, 60
17. $1,000 \times 6 = 6,000$

    $962 \times 6 = 5,772$ fish
18. $7 \times 8 = 56$ servings
19. $21 \times 100 = 2,100$ lb
20. $455 \times \$15 = \$6,825$

    $455 - 35 = 420$ people

Systematic Review 25F

1.
```
   712
  × 65
   1 1
  3550
    1
  4262
 46,280
```

2.
```
   360
  × 58
    4
  2480
   13
  1500
 20,880
```

3.
$$\begin{array}{r} 252 \\ \times 38 \\ \hline 141 \\ 1606 \\ 1 \\ 656 \\ \hline 9,576 \end{array}$$

4.
$$\begin{array}{r} 432 \\ \times \ 7 \\ \hline 121 \\ 2814 \\ \hline 3,024 \end{array}$$

5.
$$\begin{array}{r} 41 \\ \times 15 \\ \hline 205 \\ 41 \\ \hline 615 \end{array}$$

6.
$$\begin{array}{r} 62 \\ \times 23 \\ \hline 186 \\ 124 \\ \hline 1,426 \end{array}$$

7. $16 \times 38 = 608$ sq ft
8. $16 + 38 + 16 + 38 = 108$ ft
9. $10 \times 15 = 150$ sq in
10. $10 + 15 + 10 + 15 = 50$ in
11. $12 \times 12 = 144$ sq in
12. $12 \times 4 = 48$ in
13. $36 = 36$
14. $144 < 150$
15. $22 = 22$
16. 7, 14, 21, 28, 35, 42, 49, 56, 63, 70
17. $98 \times 12 = 1,176$ months
18. $11 \times 4 = 44$
    $50 - 44 = 6$ quarters
19. $400 + 200 = 600$
    $375 + 166 = 541$ copies
20. $4 \times 7 = 28$ days
    $28 \times 24 = 672$ hours

Lesson Practice 26A

1. done
2. 1×12
   2×6
   3×4
3. 1×4
   2×2
4. 1×15
   3×5
5. 1×10
   2×5
6. 1×21
   3×7
7. 1×14
   2×7
8. 1×9
   3×3
9. 1×6
   2×3
10. $25 \times 8 = 200$ pennies
11. $25 \times 12 = 300$ cents
12. $25¢ \times 3 = 75¢$
13. 1×12
    2×6
    3×4
14. 1×16
    2×8
    4×4
15. 1×18
    2×9
    3×6

Lesson Practice 26B

1. 1×16
   2×8
   4×4
2. 1×18
   2×9
   3×6
3. 22×1
   11×2

4. 1×6
   2×3
5. 1×14
   2×7
6. 1×15
   3×5
7. 1×8
   2×4
8. 1×12
   2×6
   3×4
9. 1×10
   2×5
10. 9×25 = 225 pennies
11. 16×25 = 400¢
12. 35×25 = 875¢
13. 1×20
    2×10
    4×5
14. 1×21
    3×7
15. 1×24
    2×12
    3×8
    4×6

7. 1×18
   2×9
   3×6
8. 1×9
   3×3
9. 1×4
   2×2
10. 4×25 = 100 pennies
11. 2×7 = 14
    14×25 = 350 cents
12. 5×25¢ = $1.25; yes
13. 1×16
    2×8
    4×4
14. three
    1×20
    2×10
    4×5
15. 1×12
    2×6
    3×4

Lesson Practice 26C

1. 1×22
   2×11
2. 1×8
   2×4
3. 6×1
   2×3
4. 1×14
   2×7
5. 1×21
   3×7
6. 1×10
   2×5

Systematic Review 26D

1. 1×15
   3×5
2. 1×9
   3×3
3. 1×4
   2×2
4. 11×25 = 275
5. 6×25 = 150
6. 21×25 = 525
7.   423
       ×57
        12
      2 841
        11
      2 005
     24,111

8.
```
     276
    × 12
      1
    111
    442
    276
  3,312
```

9.
```
     614
    × 32
    1228
      1
    1832
  19,648
```

10.
```
     134
    ×  4
     11
    426
    536
```

11.
```
      74
    × 33
      1
    212
      1
    212
   2442
```

12.
```
      51
    × 16
     306
      51
     816
```

13.
```
     397
    −  63
     334
```

14.
```
    3 4¹11
    −  350
       61
```

15.
```
    5⁵6 ⁹0 ¹0
    −  1 04
       4 96
```

16. $\dfrac{1}{2} = \dfrac{2}{4} = \dfrac{3}{6} = \dfrac{4}{8} = \dfrac{5}{10}$

17. 1×20
    2×10
    4×5

18. $35 \times \$22 = \$770$

19. $231 \times 2 = 462$ cars

20. $13 + 6 = 19$
    $19 \times 25 = 475$ cents or pennies

Systematic Review 26E

1. 1×6
   2×3

2. 1×12
   2×6
   3×4

3. 1×21
   3×7

4. $7 \times 25 = 175$

5. $10 \times 25 = 250$

6. $15 \times 25 = 375$

7.
```
     125
    × 54
      2
    1480
     525
   6,750
```

8.
```
     731
    × 18
      12
    5648
     731
  13,158
```

9.
```
     378
    × 49
      1
     267
    27 32
     23
   128 2
  18,5 22
```

10.
```
     276
    ×  2
      11
     442
     552
```

11.
$$\begin{array}{r} 38 \\ \times 12 \\ \hline 11 \\ 66 \\ 38 \\ \hline 456 \end{array}$$

12.
$$\begin{array}{r} 25 \\ \times 39 \\ \hline 14 \\ 185 \\ 1 \\ 65 \\ \hline 975 \end{array}$$

13.
$$\begin{array}{r} 5\,{}^4\!\!\!\!\diagup\!\!{}_5\,{}^1\!4 \\ -\quad 2\ 7 \\ \hline 5\ 2\ 7 \end{array}$$

14.
$$\begin{array}{r} 976 \\ -763 \\ \hline 2\ 13 \end{array}$$

15.
$$\begin{array}{r} {}^1\!3\,{}^{1\ 7}\!\!\!\diagup\!\!4\ 8\,{}^1\!3 \\ -\quad 2\ 9\ 9 \\ \hline 1\ 8\ 4 \end{array}$$

16. $\dfrac{2}{3} = \dfrac{4}{6} = \dfrac{6}{9} = \dfrac{8}{12} = \dfrac{10}{15}$

17. $321 \times 3 = 963$ people

18. $300 \times 3 = 900$ feet

19. two
1×15
3×5

20. $11 + 3 = 14$
$14 - 5 = 9$ times

Systematic Review 26F

1. 1×24
2×12
3×8
4×6

2. 1×16
2×8
4×4

3. 1×10
2×5

4. $2 \times 25 = 50$

5. $17 \times 25 = 425$

6. $20 \times 25 = 500$

7.
$$\begin{array}{r} 249 \\ \times 12 \\ \hline 11 \\ 488 \\ 249 \\ \hline 2{,}988 \end{array}$$

8.
$$\begin{array}{r} 218 \\ \times 75 \\ \hline 114 \\ 1050 \\ 5 \\ 1476 \\ \hline 16{,}350 \end{array}$$

9. $862 \times 10 = 8620$

10.
$$\begin{array}{r} 172 \\ \times\ 6 \\ \hline 41 \\ 622 \\ \hline 1{,}032 \end{array}$$

11.
$$\begin{array}{r} 92 \\ \times 17 \\ \hline 1 \\ 634 \\ 92 \\ \hline 1{,}564 \end{array}$$

12.
$$\begin{array}{r} 56 \\ \times 24 \\ \hline 224 \\ 112 \\ \hline 1{,}344 \end{array}$$

13.
$$\begin{array}{r} 276 \\ -\ 12 \\ \hline 264 \end{array}$$

14.
$$\begin{array}{r} {}^4\!\!\diagup\!\!{}_5\,{}^{1}4\,{}^1\!\!\diagup\!\!{}_5\,{}^1\!4 \\ -\quad 3\ 9\ 6 \\ \hline 1\ 5\ 8 \end{array}$$

15.
$$\begin{array}{r} 6\,{}^6\!\!\!\diagup\!\!7\,{}^1\!2 \\ -\ 3\ 2\ 5 \\ \hline 3\ 4\ 7 \end{array}$$

16. $\frac{5}{6} = \frac{10}{12} = \frac{15}{18} = \frac{20}{24} = \frac{25}{30}$

17. $24 \times 4 = 96$ jars

18. $12 \times 4 = 48$ quarters

19. $125 \times 45 = 5,625$ sq ft

20. $125 + 45 + 125 = 295$ ft

Lesson Practice 27A

1. done
2. two hundred sixty-one million, eight hundred twenty-nine thousand, one hundred thirty
3. done
4. 42,316
5. 149,273
6. 2,134,911
7. done
8. $100,000,000 + 50,000,000 + 900,000 + 40,000 + 1,000 + 200 + 20$
9. $600,000,000 + 400,000 + 90$
10. $6 \times 16 = 96$
11. $10 \times 16 = 160$
12. $13 \times 16 = 208$
13. $100 \times 16 = 1,600$ ounces
14. $8 \times 16 = 128$ ounces
15. $2 \times 16 = 32$
    $32 > 30$
    The two-pound can is heavier.

Lesson Practice 27B

1. sixteen million, seven hundred four thousand, nine hundred
2. three hundred twenty-one million, nine hundred fifty-four thousand
3. 4,380
4. 349,622
5. 2,461,800
6. 900,001,373

7. $10,000,000 + 1,000,000 + 600,000 + 90,000 + 1,000$
8. $500,000,000 + 9,000,000 + 400,000 + 30,000 + 2,000 + 5$
9. $400,000,000 + 50,000,000 + 1,000,000 + 600,000 + 90,000 + 8,000 + 100 + 20 + 3$
10. $9 \times 16 = 144$
11. $11 \times 16 = 176$
12. $34 \times 16 = 544$
13. $5 \times 16 = 80$ ounces
14. $7 \times 16 = 112$ ounces
15. $23 \times 16 = 368$ ounces

Lesson Practice 27C

1. three hundred eighteen million, six hundred eleven thousand, three hundred fifty-three
2. one hundred twenty-six thousand, nine hundred thirty-two
3. 23,914
4. 75,154,900
5. 6,000,342
6. 915,412,965
7. $300,000,000 + 20,000,000 + 1,000,000 + 600,000 + 10,000 + 8,000$
8. $30,000,000 + 500,000 + 800$
9. $100,000,000 + 1,000,000 + 7,000 + 3$
10. $4 \times 16 = 64$
11. $12 \times 16 = 192$
12. $51 \times 16 = 816$
13. $10 \times 16 = 160$ ounces
14. $14 \times 16 = 224$ ounces
15. $212 \times 16 = 3,392$ ounces

Systematic Review 27D

1. ten million, six hundred fifty thousand, three hundred
2. 632,178,431
3. $400,000,000 + 50,000,000 + 6,000,000 + 700,000 + 80,000 + 9,000$
4. 1×14
   2×7
5. 1×18
   2×9
   3×6
6. 1×24
   2×12
   3×8
   4×6
7. $15 \times 16 = 240$
8. $19 \times 25 = 475$
9. $9 \times 3 = 27$
10. $\begin{array}{r} 123 \\ \times 67 \\ \hline 1 \\ 1 \\ 12 \\ 741 \\ 11 \\ 628 \\ \hline 8,241 \end{array}$
11. $\begin{array}{r} 147 \\ \times 51 \\ \hline 147 \\ 2 \\ 535 \\ \hline 7,497 \end{array}$
12. $\begin{array}{r} 38 \\ \times 15 \\ \hline 14 \\ 150 \\ 38 \\ \hline 570 \end{array}$
13. $\frac{2}{8} = \frac{4}{16} = \frac{6}{24} = \frac{8}{32} = \frac{10}{40} =$
    $\frac{12}{48} = \frac{14}{56} = \frac{16}{64} = \frac{18}{72} = \frac{20}{80}$
14. $7 \times 9 = 63$ starters
15. $16 \times 4 = 64$ quarters
16. $6 + 8 = 14$
    $15 - 14 = 1$ egg
17. $30 \times 10 = 300$ miles estimated
    $29 \times 12 = 348$ miles exact
18. $5 \times 16 = 80$
    $80 + 6 = 86$ ounces

Systematic Review 27E

1. $300,000,000 + 50,000,000 + 6,000,000$
2. 784,900,000
3. four hundred million, ninety-eight
4. 1×16
   2×8
   4×4
5. 1×10
   2×5
6. 1×6
   2×3
7. $3 \times 16 = 48$
8. $20 \times 4 = 80$
9. $13 \times 3 = 39$
10. $\begin{array}{r} 557 \\ \times\ 3 \\ \hline 12 \\ 1551 \\ \hline 1,671 \end{array}$
11. $\begin{array}{r} 137 \\ \times 59 \\ \hline 1 \\ 26 \\ 973 \\ 2 \\ 13 \\ 555 \\ \hline 8,083 \end{array}$
12. $\begin{array}{r} 873 \\ \times 21 \\ \hline 11 \\ 873 \\ 1 \\ 1646 \\ \hline 18,333 \end{array}$

13. $\dfrac{3}{9} = \dfrac{6}{18} = \dfrac{9}{27} = \dfrac{12}{36} = \dfrac{15}{45} =$

$\dfrac{18}{54} = \dfrac{21}{63} = \dfrac{24}{72} = \dfrac{27}{81} = \dfrac{30}{90}$

14. $36 \times 43 = 1{,}548$ cans

15. $20 - 12 = 8$

$8 \times 16 = 128$ ounces

16. $\$200 \times 4 = \$800$

$\$175 \times 4 = \$700$

17. $\$175 \times 52 = \$9{,}100$

18. $95 + 68 + 84 + 73 + 91 = 411$ books

---

13. $\dfrac{4}{10} = \dfrac{8}{20} = \dfrac{12}{30} = \dfrac{16}{40} = \dfrac{20}{50} =$

$\dfrac{24}{60} = \dfrac{28}{70} = \dfrac{32}{80} = \dfrac{36}{90} = \dfrac{40}{100}$

14. $11 \times 13 = 143$ rolls

15. $A = 12 \times 12 = 144$ sq ft

$P = 12 + 12 + 12 + 12 = 48$ ft

16. $30 - 24 = 6$ miles

17. $5 \times 24 = 120$ cans

$12 \times 120 = 1{,}440$ ounces

18. $95 + 42 + 68 + 81 = 286$ points

Systematic Review 27F

1. $400{,}000{,}000 + 400{,}000 + 400$

2. $132{,}672{,}547$

3. six hundred ninety-eight million

4. 1×9

3×3

5. 1×15

3×5

6. 1×22

2×11

7. $20 \times 16 = 320$

8. $7 \times 8 = 56$

9. $7 \times 4 = 28$

10.
```
    412
   × 24
  1 648
  8 24
  9,888
```

11.
```
    628
   × 41
    628
   1 3
  2 482
 25,748
```

12.
```
    17
   ×52
    1
    24
   3
   55
   884
```

Lesson Practice 28A

1. done

2.
```
     284          300
   × 362        × 400
  1 1 1       120,000
    1
    468
   42
  1 284
  21
  642
 102,808
```

3.
```
     880          900
   × 153        × 200
  1   2       180,000
  2 440
  4
  4 000
   880
 1 34,640
```

4.
```
     714          700
   × 602        × 600
  1 428       420,000
  2
  4 264
 429,828
```

5.
```
   1 602        2 000
   ×    5       ×    5
   3           10,000
  5 010
  8,010
```

6.
```
    1 768      2 000
  ×    12    ×   10
    1 1 1     20,000
    1  11
    2 426
    1 7 68
    2 1, 216
```

7.
```
    8 172      8 000
  × 354      ×  400
    1 1       3,200,000
       2
    3 2 488
    3 1
    4 0 5 50
       2
    2 43 16
    2,892,888
```

8.
```
    4 6 75       5 000
  ×  2 92      ×  300
   1 2  1 1     1,500,000
       1 11
       8 240
       5 6 4
    3 6 4 35
    1 1 1
    8 2 4 0
    1, 3 6 5, 100
```

9. $120 \times 325 = 39,000$ pages

10. $1,440 \times 7 = 10,080$ minutes

Lesson Practice 28B

1.
```
     325        300
   × 213      × 200
     1 1       60,000
       1

     965
     1
     3 25
     1
     6 40
     69,225
```

2.
```
     162        200
   × 548      × 500
     1         100,000
     41
     886
     2
     4 48
    3 1
    5 00
    88,776
```

3.
```
     536        500
   × 134      × 100
     1 1        50,000
     12
     2024
     1
     1598
     5 36
     71,824
```

4.
```
     322        300
   × 725      × 700
    1 1 1 1     210,000
     1 500
       644
    1 1
    2 144
    233,450
```

5.
```
    6 424      6 000
  ×    4     ×    4
    1  1       24,000
    2 4 686
    25,696
```

6.
```
    3 445      3 000
  ×   93     ×   90
   12 1        270,000
      111
      9 225
      3 3 4
    2 7 6 6 5
    320,385
```

7.
```
    5 627      6 000
  ×  315     ×  300
   12  1       1,800,000
      3 13
      2 5 005
      5 627
      1  2
    1 5 8 6 1
    1,772,505
```

8.
$$\begin{array}{r} 3579 \\ \times 462 \\ \hline {}^1111 \\ 6048 \\ {}^2345 \\ {}^118024 \\ 223 \\ 12086 \\ \hline 1,653,498 \end{array}$$
$$\begin{array}{r} 4000 \\ \times 500 \\ \hline 2,000,000 \end{array}$$

9. $\$2,450 \times 12 = \$29,400$

10. $1,350 \times 396 = 534,600$ times

Lesson Practice 28C

1.
$$\begin{array}{r} 627 \\ \times 450 \\ \hline {}^11 \\ 13 \\ 3005 \\ 2 \\ 2488 \\ \hline 282,150 \end{array}$$
$$\begin{array}{r} 600 \\ \times 500 \\ \hline 300,000 \end{array}$$

2.
$$\begin{array}{r} 334 \\ \times 702 \\ \hline {}^1668 \\ \\ 22 \\ 2118 \\ \hline 234,468 \end{array}$$
$$\begin{array}{r} 300 \\ \times 700 \\ \hline 210,000 \end{array}$$

3.
$$\begin{array}{r} 234 \\ \times 121 \\ \hline {}^11 \\ 234 \\ 468 \\ 234 \\ \hline 28,314 \end{array}$$
$$\begin{array}{r} 200 \\ \times 100 \\ \hline 20,000 \end{array}$$

4.
$$\begin{array}{r} 415 \\ \times 378 \\ \hline 111 \\ 4 \\ 3280 \\ 3 \\ 2875 \\ 1 \\ 1235 \\ \hline 156,870 \end{array}$$
$$\begin{array}{r} 400 \\ \times 400 \\ \hline 160,000 \end{array}$$

5.
$$\begin{array}{r} 3567 \\ \times8 \\ \hline 1 \\ 445 \\ 24086 \\ \hline 28,536 \end{array}$$
$$\begin{array}{r} 4000 \\ \times8 \\ \hline 32,000 \end{array}$$

6.
$$\begin{array}{r} 2317 \\ \times64 \\ \hline {}^11 \\ 12 \\ 8248 \\ 14 \\ 12862 \\ \hline 148,288 \end{array}$$
$$\begin{array}{r} 2000 \\ \times60 \\ \hline 120,000 \end{array}$$

7.
$$\begin{array}{r} 6536 \\ \times 121 \\ \hline 11 \\ 6536 \\ 11 \\ 12062 \\ 6536 \\ \hline 790,856 \end{array}$$
$$\begin{array}{r} 7000 \\ \times 100 \\ \hline 700,000 \end{array}$$

8.
$$\begin{array}{r} 1562 \\ \times 231 \\ \hline {}^111 \\ 1562 \\ 11 \\ 3586 \\ 11 \\ 2024 \\ \hline 360,822 \end{array}$$
$$\begin{array}{r} 2000 \\ \times 200 \\ \hline 400,000 \end{array}$$

9. $3,600 \times 24 = 86,400$ seconds

10. $1,260 \times 845 = 1,064,700$ sq ft

Systematic Review 28D

1.
$$\begin{array}{r} 125 \\ \times 306 \\ \hline {}^113 \\ \\ 620 \\ 1 \\ 365 \\ \hline 38,250 \end{array}$$
$$\begin{array}{r} 100 \\ \times 300 \\ \hline 30,000 \end{array}$$

2.
```
      7 256        7 000
    ×   43       ×   40
     111 1        280,000
        11
     21 658
     2 2
     28 804
    312,008
```

3.
```
      8 761        9 000
    ×  280       ×  300
   11 1 1        2,700,000
       54   0
     6 4688
     1 1
     1 6 422
    2, 4 53,080
```

4. $100,000,000 + 20,000,000 +$
   $3,000,000 + 600,000$

5. 8×1
   4×2

6. 1×4
   2×2

7. 1×12
   2×6
   3×4

8. 1×24
   2×12
   3×8
   4×6

9. $8 \times \underline{8} = 64$

10. $7 \times \underline{6} = 42$

11. $3 \times \underline{1} = 3$

12. $9 \times \underline{5} = 45$

13. $100 + 20 = 120$ yd
    $120 \times 3 = 360$ ft

14. $3 \times 2 = 6$ lb

15. $5 \times 3 = 15$ tsp

16. $325 \times 60 = 19,500$ gallons

17. $3,572 \times \$213 = \$760,836$

18. $144 + 50 + 203 = 397$ pipers

Systematic Review 28E

1.
```
       433         400
     ×127        × 100
      11          40,000
       22
     2811
      866
     433
    54,991
```

2.
```
      8 192        8 000
    ×   74       ×   70
    111 1         560,000
        3
     32 468
     6 1
    56 734
    606,208
```

3.
```
      6 123        6 000
    ×  245       ×  200
      1 2        1,200,000
         11
     30 505
      1   1
     24 482
    12 246
    1,500,135
```

4. nine million, five hundred fifty-one thousand

5. 1×22
   2×11

6. 1×18
   2×9
   3×6

7. 1×14
   2×7

8. 1×10
   2×5

9. $9 \times \underline{9} = 81$

10. $6 \times \underline{9} = 54$

11. $4 \times \underline{5} = 20$

12. $8 \times \underline{0} = 0$

13. $125 \times 1,200 = 150,000$ pounds

14. $125 \times 16 = 2,000$ ounces

15. $43 \times 12 = 516$ stamps

16. $12 \times 15 = 180$ tiles

17. $180 \times \$2 = \$360$

18. $18+26=44$
$44-29=15$ stories

Systematic Review 28F

1.
```
    156        200
  ×523       ×500
 1  11     100,000
   358
 1 1
 2 0 2
 2 3
 5 5 0
 81,588
```

2.
```
   7481      7000
  ×  27      ×  30
    21      210,000
    25
  49867
 1  1
 14862
 201,987
```

3.
```
   1222      1000
   ×443      ×400
  1221     400,000
   3666
   4888
   4888
  541,346
```

4. 65,910,000
5. $1×9$
$3×3$
6. $1×20$
$2×10$
$4×5$
7. $1×15$
$3×5$
8. $1×21$
$3×7$
9. $7×\underline{8}=56$
10. $8×\underline{9}=72$
11. $5×\underline{10}=50$
12. $6×\underline{6}=36$
13. $13×13=169$ apples

14. $60×60=3,600$ seconds
15. $30×3=90$ bows
16. $128×275=35,200$ ounces
17. $5+2=7$ pounds
$7-3=4$ pounds
$4×16=64$ ounces
18. $100×20=2,000$ windows

Lesson Practice 29A

1. done
2. done
3. $1×9$
$3×3$; composite
4. $1×24$
$2×12$
$3×8$
$4×6$; composite
5. $1×7$; prime
6. $1×15$
$3×5$; composite
7. $1×19$; prime
8. $1×6$
$2×3$; composite
9. $5×12-60$
10. $10×12=120$
11. $2×12=24$
12. $9×12=108$
13. $4×12=48$
14. $7×12=84$
15. $12×12=144$
16. $1×12=12$
17. $8×12=96$
18. $3×12=36$ eggs
19. $6×12=72$ inches
20. $11×12=132$ months
Challenge:
12, 24, 36, 48, 60, 72, 84, 96, 108, 120, 132, 144

Lesson Practice 29B

1. 1×2; prime
2. 1×10
   2×5; composite
3. 1×17; prime
4. 1×22
   2×11; composite
5. 1×8
   2×4; composite
6. 1×3; prime
7. 1×12
   2×6
   3×4; composite
8. 1×21
   3×7; composite
9. $12 \times 12 = 144$
10. $8 \times 12 = 96$
11. $6 \times 12 = 72$
12. $3 \times 12 = 36$
13. $11 \times 12 = 132$
14. $4 \times 12 = 48$
15. $9 \times 12 = 108$
16. $2 \times 12 = 24$
17. $10 \times 12 = 120$
18. $1 \times 12 = 12$ in
19. $5 \times 12 = 60$ months
20. $7 \times 12 = 84$ people
    Challenge:
    12, 24, 36, 48, 60, 72, 84,
    96, 108, 120, 132, 144

Lesson Practice 29C

1. 1×14
   2×7; composite
2. 1×18
   2×9
   3×6; composite
3. 1×5; prime
4. 1×4
   2×2; composite
5. 1×11; prime

6. 1×20
   2×10
   4×5; composite
7. 1×12
   2×6
   3×4; composite
8. 1×23; prime
9. $2 \times 12 = 24$
10. $7 \times 12 = 84$
11. $11 \times 12 = 132$
12. $1 \times 12 = 12$
13. $6 \times 12 = 72$
14. $5 \times 12 = 60$
15. $3 \times 12 = 36$
16. $9 \times 12 = 108$
17. $12 \times 12 = 144$
18. $4 \times 12 = 48$ months
19. $10 \times 12 = 120$ pencils
20. $8 \times 12 = 96$ inches
    Challenge: 12, 24, 36, 48, 60,
    72, 84, 96, 108, 120, 132, 144

Systematic Review 29D

1. 1×6
   2×3; composite
2. 1×15
   3×5; composite
3. 1×7; prime
4. $6 \times 12 = 72$
5. $3 \times 12 = 36$
6. $8 \times 12 = 96$
7.
$$
\begin{array}{r}
45 \\
\times 33 \\
\hline
1 \\
1\ 25 \\
1 \\
1\ 25 \\
\hline
1,485
\end{array}
$$

8.
```
    4082
  ×  23
  12246
  8164
  93,886
```

9.
```
    1499
   ×770
    221
   266 0
   7833
   266
   7833
  1,154,230
```

10.
```
   8 9 1 4 3
  -  6 50
     2 93
```

11.
```
   1 3 1 2 1
  + 2 79
    6 00
```

12.
```
    476
  +8 13
  1,289
```

13. seventy-six million, eight hundred
    ninety-three thousand,
    four hundred twenty

14. 4×21 = 84 cards

15. 84×$3 = $252

16. 163×12 = 1,956 inches

17. 3,600×168 = 604,800 seconds

18. 463+584 = 1,047 sheep

Systematic Review 29E

1. 1×13; prime

2. 1×16
   2×8
   4×4; composite

3. 1×19; prime

4. 9×12 = 108

5. 12×12 = 144

6. 4×12 = 48

7.
```
     183
   ×644
    1 1
       31
      422
     31
     422
     41
     688
   117,852
```

8.
```
    8714
  ×  68
   1111
     5 3
   64682
    4 2
  48264
  592,552
```

9.
```
     2408
   ×716
    1111
      2
   12448
    2408
    2
   14856
  1,724,128
```

10.
```
    521
  ↓765
  1,286
```

11.
```
   2 0 1 4
  -1 0 8
   1 0 6
```

12.
```
   1 2 1 57
  + 4 63
    7 20
```

13. 100,000,000 + 2,000,000 +
    500,000 + 700 + 60

14. 8×10 = 80
    8×5 = 40
    80+40 = 120 pennies

15. 3,500×7 = 24,500 miles

16. 8+6+8+6 = 28 feet
    28×12 = 336 inches

17. 3×2 = 6
    6×7 = $42

18. $26 + 76 = 102$
$102 - 8 = 94$ keys

Systematic Review 29F

1. 1×10
2×5; composite
2. 1×17; prime
3. 1×22
2×11; composite
4. $11 \times 12 = 132$
5. $10 \times 12 = 120$
6. $5 \times 12 = 60$
7.
```
     964
   ×205
       1
      32
    4500
       1
    1828
  197,620
```
8.
```
    3572
   ×  12
     111
    6044
    3572
   42,864
```
9.
```
       6873
      × 343
    1 1 1 2
        22
      18419
      321
    24282
      22
    18419
  2,357,439
```
10.
```
    987
   −732
    255
```
11.
```
    1135
   +279
    4 14
```

12.
```
    1862
   + 345
    1,207
```
13. 264,510,000
14. $13 + 12 = 25$
$25 - 7 = 18$ jars
15. $4 \times 16 = 64$ ounces
16. $5 \times 4 = 20$
$19 + 1 = 20$
Enough for her students and herself.
17. $10 \times 6 = 60$
$6 \times 10 = 600$ tours
18. $1,547 \times \$5 = \$7,735$

Lesson Practice 30A

1. $1 \times 5,280 = 5,280$
2. $1 \times 2,000 = 2,000$
3. done
4. $6 \times 2,000 = 12,000$
5. $8 \times 5,280 = 42,240$
6. $11 \times 2,000 = 22,000$
7. $10 \times 5,280 = 52,800$
8. $4 \times 2,000 = 8,000$
9. $2 \times 2,000 = 4,000$ lb
10. $3 \times 2,000 = 6,000$ lb
$6,000 > 5,000$; no
11. $2 \times 5,280 = 10,560$ ft
12. $12 \times 5,280 = 63,360$ inches

Lesson Practice 30B

1. $1 \times 5,280 = 5,280$
2. $1 \times 2,000 = 2,000$
3. $9 \times 5,280 = 47,520$
4. $10 \times 2,000 = 20,000$
5. $11 \times 5,280 = 58,080$
6. $5 \times 2,000 = 10,000$
7. $16 \times 5,280 = 84,480$
8. $35 \times 2,000 = 70,000$

9. $5 \times 5,280 = 26,400$ ft
10. $22 \times 2,000 = 44,000$ lb
11. $4 \times 5,280 = 21,120$ plants
12. $7 \times 2,000 = 14,000$ lb

Lesson Practice 30C

1. $1 \times 5,280 = 5,280$
2. $1 \times 2,000 = 2,000$
3. $6 \times 5,280 = 31,680$
4. $9 \times 2,000 = 18,000$
5. $13 \times 5,280 = 68,640$
6. $12 \times 2,000 = 24,000$
7. $30 \times 5,280 = 158,400$
8. $123 \times 2,000 = 246,000$
9. $13 \times 2,000 = 26,000$ lb
10. $7 \times 2,000 = 14,000$ lb
11. $6 \times 5,280 = 31,680$ ft
    $31,680 > 30,000$
    Sarah walked further.
12. $25 \times 5,280 = 132,000$ ft

Systematic Review 30D

1. $8 \times 5,280 = 42,240$
2. $4 \times 2,000 = 8,000$
3. 3×1; prime
4. 1×21
    3×7; composite
5. 1×8
    2×4; composite
6. 1×11; prime
7. $7 \times 2 = 14$
8. $5 \times 5 = 25$
9. $10 \times 10 = 100$
10. $20 \times 3 = 60$

11.
$$
\begin{array}{r}
563 \\
\times 248 \\
\hline
1\ 1 \\
4\ 2 \\
4\ 0\ 8\ 4 \\
2\ 1 \\
2\ 0\ 4\ 2 \\
1 \\
1\ 0\ 2\ 6 \\
\hline
139,624
\end{array}
$$

12.
$$
\begin{array}{r}
8657 \\
\times\ \ 15 \\
\hline
1 \\
3\ 23 \\
4\ 0\ 0\ 5\ 5 \\
8\ 6\ 5\ 7 \\
\hline
129,855
\end{array}
$$

13.
$$
\begin{array}{r}
6214 \\
\times 572 \\
\hline
1\ 1\ 1 \\
1\ 2\ 428 \\
1\ \ \ 2 \\
4\ 2\ 478 \\
1\ \ \ 2 \\
3\ 0\ 0\ 5\ 0 \\
\hline
3,554,408
\end{array}
$$

14. $5 < 8$
15. $56 > 54$
16. $24 = 24$
17. $72 \times 2,000 = 144,000$ lb
18. $47 \times \$2 = \$94$
19. $11 \times 11 = 121$ fingers
20. $14 - 5 = 9$ feet
    $9 + 12 = 21$ feet

Systematic Review 30E

1. $10 \times 5,280 = 52,800$
2. $8 \times 2,000 = 16,000$
3. 1×4
    2×2; composite
4. 1×18
    2×9
    3×6; composite
5. 23×1; prime

6. 1×14

   2×7; composite

7. $9 \times 3 = 27$

8. $7 \times 4 = 28$

9. $4 \times 4 = 16$

10. $8 \times 8 = 64$

11.
$$
\begin{array}{r}
452 \\
\times\ 71 \\
\hline
11 \\
452 \\
31 \\
2854 \\
\hline
32{,}092
\end{array}
$$

12.
$$
\begin{array}{r}
1372 \\
\times\ 81 \\
\hline
111 \\
1372 \\
251 \\
8466 \\
\hline
111{,}132
\end{array}
$$

13.
$$
\begin{array}{r}
4912 \\
\times 131 \\
\hline
111 \\
4912 \\
2 \\
12736 \\
4912 \\
\hline
643{,}472
\end{array}
$$

14. $36 = 36$

15. $49 > 48$

16. $72 < 81$

17. $5{,}280 \times 3 = 15{,}840$ ft

    $15{,}840 < 21{,}000$; yes

18. $12 + 13 = 25$ tons

    $25 \times 2{,}000 = 50{,}000$ pounds

19. $613 \times 466 = 285{,}658$ sq ft

20. $\$68 \times 4 = \$272$

    $\$272 + \$275 = \underline{\$547}$

    $\$700 - \$547 = \underline{\$153}$

Systematic Review 30F

1. $12 \times 5{,}280 = 63{,}360$

2. $11 \times 2{,}000 = 22{,}000$

3. 1×5; prime

4. 20×1

   10×2

   4×5; composite

5. 1×16 .

   2×8

   4×4; composite

6. 1×19; prime

7. $11 \times 25 = 275$

8. $10 \times 16 = 160$

9. $5 \times 12 = 60$

10. $12 \times 12 = 144$

11.
$$
\begin{array}{r}
678 \\
\times 125 \\
\hline
111 \\
34 \\
3050 \\
11 \\
1246 \\
678 \\
\hline
84{,}750
\end{array}
$$

12.
$$
\begin{array}{r}
1563 \\
\times\ 64 \\
\hline
111 \\
221 \\
4042 \\
331 \\
6068 \\
\hline
100{,}032
\end{array}
$$

13.
$$
\begin{array}{r}
6473 \\
\times 210 \\
\hline
1\ \ 1\ \ 0 \\
6473 \\
1 \\
12846 \\
\hline
1{,}359{,}330
\end{array}
$$

14. $42 > 40$

15. $72 > 70$

16. $27 < 28$

17. $5{,}000 \times 33 = 165{,}000$ ft

18. $5 \times 2{,}000 = 10{,}000$ lb

19. $P = 3 + 3 + 3 + 3 = 12$ ft

    $A = 3 \times 3 = 9$ sq ft

20. $3 \times 12 = 36$ in

    $36 \times 36 = 1{,}296$ sq in

Test 1

1. $4 \times 3 = 12$
    $3 \times 4 = 12$
2. $5 \times 4 = 20$
    $4 \times 5 = 20$
3. $4 \times 2 = 8$
    $2 \times 4 = 8$
4. $2 \times 3 = 6$
    $3 \times 2 = 6$
5. $3 \times 1 = 3$
    $1 \times 3 = 3$
6. $5 \times 3 = 15$
    $3 \times 5 = 15$
7.
8.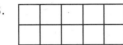

Test 2

1. $0 \times 2 = 0$
2. $9 \times 1 = 9$
3. $1 \times 7 = 7$
4. $5 \times 1 = 5$
5. $0 \times 9 = 0$
6. $3 \times 0 = 0$
7. $0 \times 7 = 0$
8. $2 \times 1 = 2$
9. $1 \times 6 = 6$
10. $1 \times 1 = 1$
11. $0 \times 1 = 0$
12. $8 \times 1 = 8$
13. $1 \times 3 = 3$
14. $6 \times 0 = 0$
15. $0 \times 5 = 0$
16. $4 \times 0 = 0$
17. $4 \times 1 = 4$
18. $8 \times 0 = 0$
19. $1 \times 9 = 9$
20. $1 \times 5 = 5$
21. $5 = 5$
22. $0 + 0 + 0 + 0 = 0$
23. $7 = 7$
24. $1 \times 6 = 6$ pieces
25. $10 \times 1 = 10$ children

Test 3

1. 2, 4, 6, 8, 10,
    12, 14, 16, 18, 20
2. 5, 10, 15, 20, 25,
    30, 35, 40, 45, 50
3. 10, 20, 30, 40, 50,
    60, 70, 80, 90, 100
4. $8 \times 1 = 8$
5. $0 \times 5 = 0$
6. $2 \times 1 = 2$
7. $9 \times 0 = 0$
8. $1 \times 7 = 7$
9. $1 \times 0 = 0$
10. $6 \times 1 = 6$
11. $4 \times 0 = 0$
12. $5 + 7 = 12$
13. $14 - 5 = 9$
14. $13 - 6 = 7$
15. $7 + 8 = 15$
16. $0 + 0 + 0 + 0 + 0 = 0$
17. $1 + 1 + 1 + 1 + 1 + 1 + 1 + 1 = 8$
18. 5, 10, 15 books
19. 10, 20, 30, 40 jelly beans
20. 2, 4, 6, 8, 10, 12, 14, 16 apples

Test 4

1. $9 \times 2 = 18$
2. $2 \times 4 = 8$
3. $1 \times 2 = 2$
4. $2 \times 3 = 6$
5. $6 \times 2 = 12$
6. $2 \times 5 = 10$
7. $7 \times 2 = 14$
8. $2 \times 2 = 4$
9. $8 \times 2 = 16$
10. $6 \times 0 = 0$
11. $10 \times 2 = 20$
12. $3 \times 1 = 3$
13. 5, 10, 15, 20, 25, 30, 35, 40, 45, 50
14. $9 - 4 = 5$
15. $5 + 3 = 8$
16. $17 - 7 = 10$
17. $9 + 8 = 17$
18. $200 + 60 + 3$
19. $2 \times 6 = 12$ jars
20. $2 \times 4 = 8$ pieces

Test 5

1. $2 \times 10 = 20$
2. $10 \times 9 = 90$
3. $3 \times 10 = 30$
4. $10 \times 7 = 70$
5. $6 \times 10 = 60$
6. $10 \times 1 = 10$
7. $4 \times 10 = 40$
8. $10 \times 5 = 50$
9. $10 \times 8 = 80$
10. $5 \times 2 = 10$
11. $1 \times 3 = 3$
12. $8 \times 2 = 16$
13. $$\begin{array}{r} 34 \\ -21 \\ \hline 13 \end{array}$$
14. $$\begin{array}{r} 55 \\ +42 \\ \hline 97 \end{array}$$

15. $18 - 1 = 17$
16. $$\begin{array}{r} 60 \\ +17 \\ \hline 77 \end{array}$$
17. $100 + 90 + 4$
18. $7 \times 10 = 70$ cents
19. $10 \times 2 = 20$ pt
    $20 - 10 = 10$ pt
20. $3 \times 10 = 30$ mi

Test 6

1. $3 \times 5 = 15$
2. $5 \times 7 = 35$
3. $2 \times 5 = 10$
4. $5 \times 8 = 40$
5. $6 \times 5 = 30$
6. $5 \times 0 = 0$
7. $10 \times 5 = 50$
8. $5 \times 5 = 25$
9. $5 \times 4 = 20$
10. $2 \times 7 = 14$
11. $9 \times 2 = 18$
12. $10 \times 6 = 60$
13. $$\begin{array}{r} 1 \\ 61 \\ +39 \\ \hline 100 \end{array}$$
14. $$\begin{array}{r} 1 \\ 47 \\ +25 \\ \hline 72 \end{array}$$
15. $$\begin{array}{r} 1 \\ 56 \\ +36 \\ \hline 92 \end{array}$$
16. $$\begin{array}{r} 1 \\ 84 \\ +19 \\ \hline 103 \end{array}$$
17. $5 + 5 + 5 + 5 + 5 + 5 + 5 + 5 = 40$
18. $2 \times 4 = 8$
    $5 \times 3 = 15$
    $8 + 15 = 23$ chores

19. $3 \times 5 = 15$¢
$8 \times 10 = 80$¢
$15 + 80 = 95$¢
20. $6 \times 5 = 30$ minutes

Unit Test I

1. $2 \times 2 = 4$
2. $7 \times 10 = 70$
3. $10 \times 10 = 100$
4. $5 \times 2 = 10$
5. $0 \times 1 = 0$
6. $10 \times 4 = 40$
7. $2 \times 8 = 16$
8. $1 \times 5 = 5$
9. $5 \times 9 = 45$
10. $4 \times 5 = 20$
11. $1 \times 6 = 6$
12. $3 \times 5 = 15$
13. $1 \times 9 = 9$
14. $10 \times 2 = 20$
15. $3 \times 10 = 30$
16. $5 \times 8 = 40$
17. $0 \times 0 = 0$
18. $10 \times 6 = 60$
19. $6 \times 5 = 30$
20. $2 \times 7 = 14$
21. $3 \times 0 = 0$
22. $2 \times 3 = 6$
23. $7 \times 5 = 35$
24. $2 \times 9 = 18$
25. $6 \times 2 = 12$
26. $10 \times 8 = 80$
27. $4 \times 2 = 8$
28. $5 \times 5 = 25$
29. $10 \times 9 = 90$
30. $0 \times 2 = 0$
31. $7 \times 1 = 7$
32. $10 \times 5 = 50$
33. 2
34. 10
35. 5

Test 7

1. $5 \times 1 = 5$ sq ft
2. $10 \times 5 = 50$ sq mi
3. $1 \times 2 = 2$ sq ft
4. $10 \times 1 = 10$ sq in
5. $5 \times 2 = 10$ sq in
6. $10 \times 2 = 20$ sq yd
7. $5 \times 6 = 30$
8. $2 \times 8 = 16$
9. $10 \times 7 = 70$
10. $5 \times 5 = 25$
11. $2 \times 4 = 8$ sq ft
12. $3 \times 2 = 6$ sq ft

Test 8

1. $10 \times \underline{2} = 20$
2. $3 \times \underline{1} = 3$
3. $9 \times \underline{5} = 45$
4. $5 \times \underline{10} = 50$
5. $2 \times \underline{4} = 8$
6. $5 \times \underline{2} = 10$
7. $2 \times \underline{9} = 18$
8. $7 \times \underline{2} = 14$
9. $7 \times 5 = 35$ sq mi
10. $10 \times 10 = 100$ sq in
11. $2 \times 3 = 6$ sq ft
12. $$\begin{array}{r} 1 \\ 39 \\ +52 \\ \hline 91 \end{array}$$
13. $$\begin{array}{r} {}^8 9 \,{}^1 5 \\ -\ 1\ 6 \\ \hline 7\ 9 \end{array}$$
14. $$\begin{array}{r} 1 \\ 78 \\ +25 \\ \hline 103 \end{array}$$

15.
$$\begin{array}{r} {}^{5}\!\!\!\not{6}\; {}^{1}\!\!\!1 \\ -\;4\;8 \\ \hline 1\;3 \end{array}$$

16. $45 - 37 = 8$ mi

17. $5 \times \underline{8} = 40$; 8 trips

18. $2 \times \underline{6} = 12$; 6 qt

19. $\$5 \times \underline{7} = \$35$; 7 five-dollar bills

20. $10 \times \underline{5} = 50$; 5 steps

14.
$$\begin{array}{r} {}^{6}\!\!\!\not{7}\; {}^{1}\!\!\!0 \\ -\;2\;1 \\ \hline 4\;9 \end{array}$$

15. $4 \times 10¢ = 40¢$

$3 \times 5¢ = 15¢$

$40¢ + 15¢ = 55¢$

16. 9, 18, <u>27</u> marbles

17. $5 \times \$9 = \$45$

18. 9, 18, 27, 36, 45, 54, 63, 72, <u>81</u> chips

Test 9

1. 9, 18, 27, 36, 45,
54, 63, 72, 81, 90

2. $\dfrac{2}{9} = \dfrac{4}{18} = \dfrac{6}{27} = \dfrac{8}{36} = \dfrac{10}{45} =$

$\dfrac{12}{54} = \dfrac{14}{63} = \dfrac{16}{72} = \dfrac{18}{81} = \dfrac{20}{90}$

3. $\dfrac{0}{(5)(0)} , \dfrac{5}{(5)(1)} , \dfrac{10}{(5)(2)} , \dfrac{15}{(5)(3)} ,$

$\dfrac{20}{(5)(4)} , \dfrac{25}{(5)(5)} , \dfrac{30}{(5)(6)} , \dfrac{35}{(5)(7)} ,$

$\dfrac{40}{(5)(8)} , \dfrac{45}{(5)(9)} , \dfrac{50}{(5)(10)}$

4. $5 \times \underline{9} = 45$

5. $8 \times \underline{2} = 16$

6. $5 \times \underline{7} = 35$

7. $8 \times \underline{10} = 80$

8. $8 \times 1 = 8$ sq mi

9. $1 \times 1 = 1$ sq in

10. $6 \times 5 = 30$ sq ft

11.
$$\begin{array}{r} 1 \\ 32 \\ +58 \\ \hline 90 \end{array}$$

12.
$$\begin{array}{r} {}^{4}\!\!\!\not{5}\; {}^{1}\!\!\!4 \\ -\;4\;7 \\ \hline 7 \end{array}$$

13.
$$\begin{array}{r} 1 \\ 65 \\ +18 \\ \hline 83 \end{array}$$

Test 10

1. $9 \times 0 = 0$

2. $9 \times 9 = 81$

3. $3 \times 9 = 27$

4. $7 \times 9 = 63$

5. $9 \times 4 = 36$

6. $6 \times 9 = 54$

7. $9 \times 8 = 72$

8. $2 \times 9 = 18$

9. $10 \times 9 = 90$

10. $9 \times 5 = 45$

11. $2 \times 6 = 12$

12. $5 \times 8 = 40$

13. $\dfrac{0}{9 \bullet 0} , \dfrac{9}{9 \bullet 1} , \dfrac{18}{9 \bullet 2} , \dfrac{27}{9 \bullet 3} , \dfrac{36}{9 \bullet 4} , \dfrac{45}{9 \bullet 5}$

$\dfrac{54}{9 \bullet 6} , \dfrac{63}{9 \bullet 7} , \dfrac{72}{9 \bullet 8} , \dfrac{81}{9 \bullet 9} , \dfrac{90}{9 \bullet 10}$

14. $2 \times \underline{9} = 18$

15. $5 \times \underline{5} = 25$

16. $4 \times \underline{0} = 0$

17. $8 \times \underline{9} = 72$

18. $9 \times 3 = 27$ minutes

19. $9 \times 9 = 81$ lives

20. $\$9 \times 8 = \$72$

Test 11

1. 3, 6, 9, 12, 15, 18, 21, 24, 27, 30

2. $\dfrac{3}{5} = \dfrac{6}{10} = \dfrac{9}{15} = \dfrac{12}{20} = \dfrac{15}{25} =$

$\dfrac{18}{30} = \dfrac{21}{35} = \dfrac{24}{40} = \dfrac{27}{45} = \dfrac{30}{50}$

3. $\dfrac{0}{9\times0}, \dfrac{9}{9\times1}, \dfrac{18}{9\times2}, \dfrac{27}{9\times3}, \dfrac{36}{9\times4}, \dfrac{45}{9\times5},$

$\dfrac{54}{9\times6}, \dfrac{63}{9\times7}, \dfrac{72}{9\times8}, \dfrac{81}{9\times9}, \dfrac{90}{9\times10}$

4. $5 \times 4 = 20$

5. $2 \times 7 = 14$

6. $5 \times 3 = 15$

7. $2 \times 8 = 16$

8.
```
  1
  21
  39
+  6
  66
```

9.
```
  28
  40
 +61
 129
```

10.
```
  2
  65
  23
  15
 + 7
 110
```

11.
```
  1
  32
  33
  42
 +13
 120
```

12. 3, 6, 9, <u>12</u> ribbons

13. 3, 6, 9, 12, 15, 18, <u>21</u> meals

14. 3, 6, 9, 12, <u>15</u> pages

15. 3, 6, 9, 12, 15, <u>18</u> miles

Test 12

1. $3 \times 9 = 27$

2. $8 \times 3 = 24$

3. $3 \times 3 = 9$

4. $4 \times 3 = 12$

5. $3 \times 2 = 6$

6. $6 \times 3 = 18$

7. $3 \times 10 = 30$

8. $7 \times 3 = 21$

9. $3 \times 1 = 3$

10. $5 \times 3 = 15$

11. $6 \times 9 = 54$

12. $9 \times 9 = 81$

13.
```
  1
  42
  34
 + 8
  84
```

14.
```
  1
  17
  11
 +13
  41
```

15.
```
  8 1
  9 2
 -2 5
  6 7
```

16.
```
  7 1
  8 4
 -3 6
  4 8
```

17. $6 \times 3 = 18'$

18. $9 \times 3 = 27$

19. $9 \times 3 = 27'$

20. $3 \times 4 = 12$ sandwiches

Test 13

1. 6, 12, 18, 24, 30, 36, 42, 48, 54, 60

2. 9, 18, 27, 36, 45, 54, 63, 72, 81, 90

3. $\dfrac{1}{2} = \dfrac{2}{4} = \dfrac{3}{6} = \dfrac{4}{8} = \dfrac{5}{10}$

4. $3 \times 3 = 9$

5. $3 \times 9 = 27$

6. $9 \times 9 = 81$

7. $0 \times 0 = 0$

8. $10 + 10 + 13 + 13 = 46$ mi

9. $15 + 32 + 15 + 32 = 94'$

10. $6 + 6 + 6 + 6 = 24''$

11. 6, 12, 18, 24, <u>30</u> bulbs

12. 6, 12, 18, 24, 30, 36, <u>42</u> letters

13. 6, 12, 18, <u>24</u> blocks

14. 6, 12, 18, 24, 30, <u>36</u> trips

Test 14

1. $6 \times 7 = 42$
2. $9 \times 6 = 54$
3. $6 \times 3 = 18$
4. $1 \times 6 = 6$
5. $6 \times 4 = 24$
6. $10 \times 6 = 60$
7. $6 \times 8 = 48$
8. $6 \times 6 = 36$
9. $6 \times 2 = 12$
10. $3 \times 3 = 9$
11. $6 \times 5 = 30$
12. $2 \times 10 = 20$
13. $5 \times 5 = 25$
14. $7 \times 9 = 63$
15. $9 \times 9 = 81$
16. $8 \times 3 = 24$
17. $\dfrac{5}{6} = \dfrac{10}{12} = \dfrac{15}{18} = \dfrac{20}{24} = \dfrac{25}{30}$
18. $2 \times 3 = 6$ tsp
19. $6 \times 3 = 18$ ft
20. $6 \times 6 = 36$ cups

Unit Test II

1. $8 \times 6 = 48$
2. $3 \times 5 = 15$
3. $9 \times 6 = 54$
4. $3 \times 7 = 21$
5. $1 \times 9 = 9$
6. $6 \times 0 = 0$
7. $10 \times 9 = 90$
8. $9 \times 8 = 72$
9. $3 \times 2 = 6$
10. $7 \times 9 = 63$
11. $3 \times 1 = 3$
12. $3 \times 6 = 18$
13. $10 \times 3 = 30$
14. $6 \times 3 = 18$
15. $4 \times 9 = 36$
16. $6 \times 4 = 24$

17. $2 \times 6 = 12$
18. $3 \times 3 = 9$
19. $5 \times 6 = 30$
20. $9 \times 2 = 18$
21. $3 \times 9 = 27$
22. $9 \times 9 = 81$
23. $5 \times 9 = 45$
24. $3 \times 8 = 24$
25. $4 \times 3 = 12$
26. $6 \times 6 = 36$
27. $7 \times 6 = 42$
28. $10 \times 6 = 60$
29. $2 \times 9 = 18$ sq in
30. $2 + 9 + 2 + 9 = 22''$
31. $4 \times 3 = 12'$
32. $6 \times 3 = 18$ tsp

Test 15

1. 4, 8, 12, 16, 20, 24, 28, 32, 36, 40
2. 3, 6, 9, 12, 15, 18, 21, 24, 27, 30
3. $\dfrac{2}{4} = \dfrac{4}{8} = \dfrac{6}{12} = \dfrac{8}{16} = \dfrac{10}{20}$
4. $3 \times 4 = 12$
5. $5 \times 6 = 30$
6. $9 \times 9 = 81$
7. $6 \times 7 = 42$
8. $5 \times 2 = 10$
9. $1 \times 8 = 8$
10. $9 \times 7 = 63$
11. $3 \times 3 = 9$
12. 4, 8, 12, 16, 20, <u>24</u> qt
13. 4, 8, <u>12</u> minutes
14. 4, 8, 12, <u>16</u> times
15. 4, 8, 12, 16, 20, 24, <u>28</u> hooves

Test 16

1. $4 \times 9 = 36$
2. $4 \times 4 = 16$

3. $4 \times 2 = 8$
4. $5 \times 4 = 20$
5. $4 \times 3 = 12$
6. $8 \times 4 = 32$
7. $4 \times 7 = 28$
8. $4 \times 6 = 24$
9. $10 \times 4 = 40$
10. $9 \times 6 = 54$
11. $3 \times 7 = 21$
12. $6 \times 6 = 36$
13. $5 \times 5 = 25$ sq in
14. $5+5+5+5 = 20$ in
15. $2 \times 6 = 12$
16. $3 \times 3 = 9$
17. $10 \times 3 = 30$
18. $6 \times 4 = 24$ quarters
19. $5 \times 4 = 20$ plants
20. $3 \times 4 = 12$ gum balls

Test 17

1. 7, 14, 21, 28, 35, 42, 49, 56, 63, 70
2. $\frac{6}{7} = \frac{12}{14} = \frac{18}{21} = \frac{24}{28} = \frac{30}{35}$
3. $70 \times 3 = 210$
4. $40 \times 2 = 80$
5. $90 \times 5 = 450$
6. $7 \times 2 = 14$
7. $4 \times 9 = 36$
8. $6 \times 7 = 42$
9. $5 \times 3 = 15$
10. $27 < 28$
11. $13 = 13$
12. 4 ft < 30 ft
13. 7, 14, 21, 28, 35, 42 days
14. 7, 14, 21, 28, 35, 42, 49 points
15. 7, 14, 21 pieces

Test 18

1. $7 \times 2 = 14$
2. $4 \times 7 = 28$
3. $3 \times 4 = 12$
4. $6 \times 7 = 42$
5. $7 \times 9 = 63$
6. $10 \times 7 = 70$
7. $7 \times 7 = 49$
8. $5 \times 7 = 35$
9. $8 \times 7 = 56$
10. $6 \times 6 = 36$
11. $50 \times 5 = 250$
12. $90 \times 2 = 180$
13. $100 \times 4 = 400$
14. $400 \times 2 = 800$
15. $200 \times 2 = 400$
16. $100 \times 6 = 600$
17. $4 \times 7 = 28$ quarters
18. $\$8 \times 7 = \$56$
19. $7 \times 6 = 42'$
20. $7 \times 100 = 700$ paper clips

Test 19

1. 8, 16, 24, 32, 40, 48, 56, 64, 72, 80
2. 7, 14, 21, 28, 35, 42, 49, 56, 63, 70
3. $\frac{1}{5} = \frac{2}{10} = \frac{3}{15} = \frac{4}{20} = \frac{5}{25}$
4. $7 \times 7 = 49$
5. $5 \times 7 = 35$
6. $6 \times 5 = 30$
7. $7 \times 8 = 56$
8. $8 \times 3 = 24$
9. $0 \times 9 = 0$
10. $6 \times 7 = 42$
11. $6 \times 8 = 48$
12. $\begin{array}{r} 450 \\ + 106 \\ \hline 556 \end{array}$
13. $\begin{array}{r} 1 \\ 392 \\ + 117 \\ \hline 509 \end{array}$

14.
$$\begin{array}{r} ^1 \\ 546 \\ +237 \\ \hline 783 \end{array}$$

15. 8, 16, 24, 32, 40, <u>48</u> pints

16. 8, 16, 24, 32, <u>40</u> hours

17. 8, 16, 24, 32, 40, <u>48</u> sq ft

18. 8, 16, 24, <u>32</u> quarters

Test 20

1. $8 \times 3 = 24$
2. $9 \times 8 = 72$
3. $8 \times 7 = 56$
4. $4 \times 8 = 32$
5. $8 \times 8 = 64$
6. $6 \times 8 = 48$
7. $8 \times 10 = 80$
8. $5 \times 8 = 40$
9. $80 \times 2 = 160$
10. $7 \times 6 = 42$
11. $50 \times 7 = 350$
12. $300 \times 3 = 900$
13. $9 > 8$
14. $2 \text{ pt} < 16 \text{ pt}$
15. $24 = 24$

16.
$$\begin{array}{r} 6\,\,{}^1 2\,\,{}^1 0 \\ -\,\,4\,\,3\,\,1 \\ \hline 2\,\,8\,\,9 \end{array}$$

17.
$$\begin{array}{r} 7\,\,{}^1 8\,\,{}^8 9\,\,{}^1 2 \\ -\,\,6\,\,9\,\,5 \\ \hline 1\,\,9\,\,7 \end{array}$$

18.
$$\begin{array}{r} 3\,\,{}^9 4\,\,{}^1 0\,\,{}^1 6 \\ -\,\,1\,\,1\,\,9 \\ \hline 2\,\,8\,\,7 \end{array}$$

19. $3 \times 8 = 24$ glasses
20. $8 \times 5 = 40$ words

Unit Test III

1. $4 \times 7 = 28$
2. $5 \times 5 = 25$
3. $0 \times 1 = 0$
4. $8 \times 9 = 72$
5. $3 \times 4 = 12$
6. $3 \times 8 = 24$
7. $5 \times 3 = 15$
8. $7 \times 10 = 70$
9. $3 \times 3 = 9$
10. $10 \times 6 = 60$
11. $8 \times 5 = 40$
12. $6 \times 4 = 24$
13. $8 \times 2 = 16$
14. $4 \times 9 = 36$
15. $7 \times 7 = 49$
16. $4 \times 5 = 20$
17. $6 \times 1 = 6$
18. $7 \times 5 = 35$
19. $3 \times 6 = 18$
20. $10 \times 2 = 20$
21. $9 \times 3 = 27$
22. $10 \times 5 = 50$
23. $7 \times 2 = 14$
24. $7 \times 8 = 56$
25. $10 \times 3 = 30$
26. $4 \times 2 = 8$
27. $8 \times 8 = 64$
28. $10 \times 9 = 90$
29. $5 \times 2 = 10$
30. $4 \times 4 = 16$
31. $8 \times 6 = 48$
32. $5 \times 0 = 0$
33. $10 \times 10 = 100$
34. $6 \times 9 = 54$
35. $10 \times 8 = 80$
36. $9 \times 9 = 81$
37. $6 \times 5 = 30$
38. $8 \times 4 = 32$
39. $9 \times 5 = 45$
40. $6 \times 6 = 36$
41. $2 \times 9 = 18$
42. $10 \times 4 = 40$

43. $6 \times 7 = 42$
44. $2 \times 3 = 6$
45. $6 \times 2 = 12$
46. $9 \times 7 = 63$
47. $2 \times 2 = 4$
48. $7 \times 3 = 21$
49. $5 \times 2 = 10$
50. $8 \times 3 = 24$
51. $6 \times 4 = 24$
52. $10 \times 4 = 40$
53. $4 \times 3 = 12$
54. $7 \times 8 = 56$
55. $6 \times 5 = 30$ sq ft
56. $6 + 5 + 6 + 5 = 22'$

10.
$$\begin{array}{r} {}^{1}42 \\ 19 \\ + 37 \\ \hline 98 \end{array}$$

11.
$$\begin{array}{r} 2\,{}^{3}\!4\,{}^{1}5 \\ - \quad 16 \\ \hline 2\quad 29 \end{array}$$

12.
$$\begin{array}{r} 2\,{}^{9}\!3\,{}^{1}0\,{}^{1}4 \\ - 2\quad 28 \\ \hline 76 \end{array}$$

13. $103 \times 2 = 206$ ears
14. $12 \times 4 = 48$ eggs
15. $13 \times 3 = 39$ miles

Test 21

1. $\begin{array}{r}22\\\times 4\\\hline 88\end{array}$ $\quad \begin{array}{r}20+2\\\times \quad 4\\\hline 80+8\end{array}$

2. $\begin{array}{r}12\\\times 3\\\hline 36\end{array}$ $\quad \begin{array}{r}10+2\\\times \quad 3\\\hline 30+6\end{array}$

3. $\begin{array}{r}11\\\times 5\\\hline 55\end{array}$ $\quad \begin{array}{r}10+1\\\times \quad 5\\\hline 50+5\end{array}$

4. $\begin{array}{r}41\\\times 2\\\hline 82\end{array}$ $\quad \begin{array}{r}40+1\\\times \quad 2\\\hline 80+2\end{array}$

5. $\begin{array}{r}211\\\times 3\\\hline 633\end{array}$ $\quad \begin{array}{r}200+10+1\\\times \quad 3\\\hline 600+30+3\end{array}$

6. $\begin{array}{r}202\\\times 4\\\hline 808\end{array}$ $\quad \begin{array}{r}200+00+2\\\times \quad 4\\\hline 800+00+8\end{array}$

7. $\begin{array}{r}111\\\times 7\\\hline 777\end{array}$ $\quad \begin{array}{r}100+10+1\\\times \quad 7\\\hline 700+70+7\end{array}$

8. $\begin{array}{r}112\\\times 4\\\hline 448\end{array}$ $\quad \begin{array}{r}100+10+2\\\times \quad 4\\\hline 400+40+8\end{array}$

9. $\begin{array}{r}{}^{1}2\,1\\35\\+ 24\\\hline 80\end{array}$

Test 22

1. 30
2. 40
3. 80
4. 300
5. 500
6. 100
7. 7,000
8. 3,000
9. 8,000
10. $40 \times 3 = 120$
11. $70 \times 2 = 140$
12. $30 \times 4 = 120$
13. $200 \times 7 = 1,400$
14. $800 \times 4 = 3,200$
15. $200 \times 4 = 800$
16. $\begin{array}{r}24\\\times 2\\\hline 48\end{array}$ $\quad \begin{array}{r}20+4\\\times \quad 2\\\hline 40+8\end{array}$
17. $\begin{array}{r}12\\\times 3\\\hline 36\end{array}$ $\quad \begin{array}{r}10+2\\\times \quad 3\\\hline 30+6\end{array}$
18. $2,000
19. $50 \times 6 = \$300$
20. $200 \times 4 = 800$ miles

Test 23

1.
```
    23
   ×23
   169
    46
   529
```

2.
```
    22
   ×44
   188
    88
   968
```

3.
```
    21
   ×11
    21
    21
   231
```

4.
```
    32
   ×12
    64
    32
   384
```

5.
```
    17
   ×11
    17
    17
   187
```

6.
```
    18
   ×11
    18
    18
   198
```

7.
```
    23
   ×32
   146
    69
   736
```

8.
```
    20
   ×44
    80
    80
   880
```

9. $800 \times 5 = 4,000$
10. $500 \times 3 = 1,500$
11. $700 \times 4 = 2,800$
12.
```
     11
    138
   +274
    412
```

13.
```
    705
   +398
   1103
```

14.
```
      1
    464
   +140
    604
```

15. 40
16. 4,000
17. $11 \times 14 = 154$ sq ft
18. $2 \times 7 = 14$ days
$14 \times 12 = 168$ donuts

Test 24

1.
```
    26
   ×15
   130
    26
   390
```

2.
```
    31
   ×29
   279
    62
   899
```

3.
```
    22
   ×35
   110
    66
   770
```

4.
```
     38
    ×34
      3
    122
    2
     94
   1292
```

5.
```
     28
    ×39
    17
    182
    2
     64
   1092
```

6.
```
    14
   ×16
     2
   164
    14
   224
```

7.
```
    27
   ×23
    12
   161
    44
   621
```

8.
```
    16
   ×24
     2
   144
    22
   384
```

9. $100×3 = 300$
10. $700×8 = 5,600$
11. $200×4 = 800$
12. 4, 8, 12, 16, 20, 24, 28, 32, 36, 40
13. $10×5 = 50$
14. $4×8 = 32$
15. $3×4 = 12$
16. $5×48 = 240$ miles
17. $24×36 = 864$ sq ft
18. $24×31 = 744$ hours

Test 25

1.
```
   312
   ×53
   936
    11
  1550
 16,536
```

2.
```
   258
   ×78
     1
   246
  1604
   135
  1456
 20,124
```

3.
```
    475
    ×76
      1
    243
   2420
    143
   2895
  36,100
```

4.
```
    316
    × 5
      3
   1550
  1,580
```

5.
```
     34
    ×96
     12
    184
     13
    276
   3,264
```

6.
```
     81
    ×11
     81
     81
    891
```

7. $12×35 = 420$ sq ft
8. $35+12+35+12 = 94'$
9. $15×18 = 270$ sq in
10. $15+18+15+18 = 66"$
11. $14 > 9$
12. $36 = 36$
13. $45 > 40$
14. 6, 12, 18, 24, 30, 36, 42, 48, 54, 60
15. $36×108 = 3,888$ sq in
16. $52×\$250 = \$13,000$
17. $625×55 = 34,375$ lb
18. $137×31 = 4,247$ newspapers

Test 26

1. $1×6$
   $2×3$
2. $1×20$
   $2×10$
   $4×5$

3. 1×18
   2×9
   3×6
4. 1×10
   2×5
5. 1×15
   3×5
6. 1×12
   2×6
   3×4
7. 1×9
   3×3
8. 1×14
   2×7
9. 3×25 = 75
10. 14×25 = 350
11.
$$\begin{array}{r} 185 \\ \times 13 \\ \hline 1 \\ 21 \\ 1345 \\ 185 \\ \hline 2,405 \end{array}$$
12.
$$\begin{array}{r} 693 \\ \times 21 \\ \hline 1 \\ 1693 \\ 1 \\ 1286 \\ \hline 14,553 \end{array}$$
13.
$$\begin{array}{r} 564 \\ \times 48 \\ \hline 1 \\ 143 \\ 4082 \\ 21 \\ 2046 \\ \hline 27,072 \end{array}$$
14. $\frac{3}{6} = \frac{6}{12} = \frac{9}{18} = \frac{12}{24} = \frac{15}{30}$
15. 4×6 = 24 units
16. 23×25 = 575 cents
17. 1×16
    2×8 ✓
    4×4
18. 3×25 = 75¢
    75¢ > 69¢; yes

Test 27

1. six million, seven hundred one thousand, four hundred thirteen
2. five hundred seventy thousand, three hundred forty-eight
3. 402,519
4. 179,457,385
5. 204,213
6. 900,000,000 + 70,000,000 + 5,000,000 + 200,000 + 30,000 + 6,000 + 700 + 50 + 9
7. 300,000 + 40,000 + 2,000 + 700 + 70 + 6
8. 19×16 = 304 oz
9. 47×16 = 752 oz
10. 4×16 = 64 oz
11. 1×16
    2×8
    4×4
12. 1×20
    2×10
    4×5
13. 8×1
    2×4
14. 1×6
    2×3
15. $\frac{2}{9} = \frac{4}{18} = \frac{6}{27} = \frac{8}{36} = \frac{10}{45} =$
    $\frac{12}{54} = \frac{14}{63} = \frac{16}{72} = \frac{18}{81} = \frac{20}{90}$
16. 20×16 = 320 oz   or 19×16 = 304 oz
17. 36×16 = 576 oz   or 47×16 = 752 oz
18. 26×16 = 416 oz   or 4×16 = 64 oz

Test 28

1.
$$\begin{array}{r} 524 \\ \times 566 \\ \hline 12 \\ 3024 \\ 12 \\ 3024 \\ 12 \\ 2500 \\ \hline 296,584 \end{array}$$

2.
```
      8867
    ×   93
      1 1
     1212
    24481
    1756
    72243
   824,631
```

3.
```
      4461
    × 365
        1
       23
     ¹20005
      123
     24466
      11
    12283
  1,628,265
```

4. 5,137,213

5. forty-four million,
   nine hundred thousand

6. 500,000,000+10,000,000+7,000,000+
   50,000+8,000+800

7. 1×12
   2×6
   3×4

8. 1×4
   2×2

9. 1×24
   2×12
   3×8
   4×6

10. 1×18
    2×9
    3×6

11. 6×8 = 48

12. 9×9 = 81

13. 7×7 = 49

14. 8×7 = 56

15. 125×125 = 15,625 sq ft

16. 1 widget per sec = 60 per min
    60×60 = 3,600 per hour
    3,600×8 = 28,800 widgets

Test 29

1. 1×6
   2×3; composite

2. 1×18
   2×9
   3×6; composite

3. 1×23; prime

4. 1×15
   3×5; composite

5. 1×19; prime

6. 1×12
   2×6
   3×4; composite

7. 10×12 = 120

8. 9×12 = 108

9. 6×12 = 72

10.
```
      485
    ×712
        1
       11
     2860
     1485
     153
     2865
   345,320
```

11.
```
     5491
    ×  36
        1
       25
    30446
       12
    15273
   197,676
```

12.
```
      7123
    × 547
      1212
     49741
      1  1
     28482
      11
    35505
  3,896,281
```

13.
```
       8 9¹2
    -  1 73
       7  19
```

14.
```
   11
  925
+  86
1,011
```

15.
```
  1
  204
+139
  343
```

16. 1×24

    2×12

    3×8

    4×6

17. 7×12 = 84 months

18. 6×12 = 72 eggs

13.
```
   4705
 ×   25
    3 2
  20505
   1 1
   8400
117,625
```

14. 64 > 63

15. 12 = 12

16. 25 < 28

17. 18 + 22 = 40 tons

    40 × 2,000 = 80,000 lb

18. 6 × 5,280 = 31,680 ft

19. 5 × 2,000 = 10,000 lb

20. 14 × 5,280 = 73,920 ft

Test 30

1. 5,280

2. 2,000

3. 3 × 5,280 = 15,840

4. 22 × 2,000 = 44,000

5. 7 × 5,280 = 36,960

6. 3 × 2,000 = 6,000

7. 12 × 25 = 300

8. 9 × 16 = 144

9. 11 × 12 = 132

10. 7 × 12 = 84

11.
```
   392
 ×234
    1
   13
 1268
    2
 2976
    1
  684
91,728
```

12.
```
  2638
 ×   9
    1
 1527
18472
23,742
```

Unit Test IV

1.
```
   21
 ×48
    1
  168
   84
1,008
```

2.
```
   364
 × 53
    11
  1982
    32
  1500
19,292
```

3.
```
    106
 × 789
    15
   904
    14
   808
   14
   702
83,634
```

4.
```
  1357
 ×   6
    1
  134
 6802
 8,142
```

5.
```
    2843
  ×   75
       1
    1421
   10005
   1522
   14681
  213,225
```

6.
```
    4561
  ×   32
     111
    8022
      11
   12583
  145,952
```

7. 1×9

3×3; composite

8. 1×12

2×6

3×4; composite

9. 1×7; prime

10. 8×25 = 200

11. 10×16 = 160

12. 3×12 = 36

13. 6×5,280 = 31,680

14. 2×2,000 = 4,000

15. 25×12 = 300

16. 90

17. 100

18. 5,000

19. 4,568

20. 2,000,000+300,000+
90,000+1,000+600

Final Test

1.
```
     85
   ×26
     13
    480
     11
    160
  2,210
```

2.
```
     421
   ×73
       1
    1263
    11
   2847
  30,733
```

3.
```
     509
   ×636
      15
    3004
    1 2
    1507
      5
    3004
  323,724
```

4.
```
    7546
  ×    8
    1434
   56028
   60,368
```

5.
```
    3482
  ×   59
       1
    1371
   27628
   1241
   15000
  205,438
```

6.
```
    6187
  ×467
       1
    1354
   42769
   1 44
   36682
    32
   24428
  2,889,329
```

7. 31×72 = 2,232 sq ft

8. 31+72+31+72 = 206'

9. 8×8 = 64

10. 9×7 = 63

11. 10×10 = 100

12. 1×16

2×8

4×4; composite

13. 1×7; prime

14. 1×9

3×3; composite

15. $12 = 12$

16. $72 > 60$

17. $42 < 45$

18.
$$
\begin{array}{r}
1 \\
92 \\
21 \\
48 \\
+17 \\
\hline
178
\end{array}
$$

19.
$$
\begin{array}{r}
1 \\
163 \\
+54 \\
\hline
217
\end{array}
$$

20.
$$
\begin{array}{r}
815 \\
+482 \\
\hline
1{,}297
\end{array}
$$

21.
$$
\begin{array}{r}
3\,{}^5\!6\,{}^1\!0 \\
-\quad 37 \\
\hline
3\quad 2\;3
\end{array}
$$

22.
$$
\begin{array}{r}
{}^4\!5\,{}^1\!29 \\
-\quad 168 \\
\hline
3\;6\;1
\end{array}
$$

23.
$$
\begin{array}{r}
{}^3\!4\,{}^9\!0\,{}^1\!2 \\
-\quad 2\;93 \\
\hline
1\quad 0\;9
\end{array}
$$

24. $6 \times 2 = 12$

25. $8 \times 10 = 80$

26. $9 \times 3 = 27$

27. $5 \times 3 = 15$

28. $10 \times 5 = 50$

29. $7 \times 4 = 28$

30. $2 \times 4 = 8$

31. $4 \times 8 = 32$

32. $3 \times 16 = 48$

33. $6 \times 25 = 150$

34. $2 \times 5{,}280 = 10{,}560$

35. $1 \times 2{,}000 = 2{,}000$

36. $20 \times 40 = 800$ sq ft

37. $500 \times 3 = 1{,}500$ mi

38. $3{,}000$

39. $1{,}271{,}028$

40. $5{,}000{,}000 + 600{,}000 + 80{,}000 + 1{,}000 + 900$

Symbols & Tables

SYMBOLS

=	equals
+	plus
−	minus
X	times/multiply
•	times
( )( )	times
¢	cents
$	dollars
'	foot
"	inch
<	less than
>	greater than

TIME

60 seconds = 1 minute

60 minutes = 1 hour

1 week = 7 days

1 year = 365 days

1 year = 52 weeks

1 year = 12 months

1 decade = 10 years

1 century = 100 years

MONEY

1 penny = 1 cent (1¢)

1 nickel = 5 cents (5¢)

1 dime = 10 cents (10¢)

1 quarter = 25 cents (25¢)

1 dollar = 100 cents (100¢ or $1.00)

1 dollar = 4 quarters

PLACE-VALUE NOTATION

931,452 = 900,000 + 30,000 + 1,000 + 400 + 50 + 2

LABELS FOR PARTS OF PROBLEMS

Addition

25	addend
+16	addend
41	sum

Multiplication

3 3	multiplicand
× 5	multiplier
1 6 5	product

Subtraction

4 5	minuend
− 2 2	subtrahend
2 3	difference

MEASUREMENT

1 quart (qt) = 2 pints (pt)

1 gallon (gal) = 8 pints (pt)

1 gallon (gal) = 4 quarts (qt)

1 tablespoon (Tbsp) = 3 teaspoons (tsp)

1 foot (ft) = 12 inches (in)

1 yard (yd) = 3 feet (ft)

1 mile (mi) = 5,280 feet (ft)

1 pound (lb) = 16 ounces (oz)

1 ton = 2,000 pounds (lb)

1 dozen = 12

Glossary

A-C

Area - the number of square units in a rectangle, found by multiplying the factors, or dimensions

Borrowing - see Regrouping

Carrying - see Regrouping

Century - one hundred years

Commutative property - the order of factors in a multiplication problem may be changed without changing the product. The commutative property also applies to addition.

Composite number - a number with more than one set of different factors

D-E

Decade - 10 years

Denominator - the bottom number in a fraction. It tells how many total parts there are in the whole.

Dimension - the length of one of the sides of a rectangle or other shape

Distributive property - in this book, used to show how a factor is multiplied by each different place value in multiple-digit problems

Equation - a number sentence in which the value of one side is equal to the value of the other side

Estimation - used to get an approximate value of an answer

Even number - a number that ends in 0, 2, 4, 6, or 8. Even numbers are multiples of two.

F-I

Factoring - finding the factors when the product or area is known

Factors - the two sides of a rectangle or the numbers multiplied in a multiplication problem

Fraction - one number written over another to show part of a whole. A fraction can also indicate division.

Hexagon - a shape with six sides

Inequality - a number sentence in which the value of one side is greater than the value of the other side

J-O

Multiplicand - the top factor in a multiplication problem

Multiplier - the bottom factor in a multiplication problem

Nineovers - a method used to check addition and multiplication problems

Numerator - the top number in a fraction. It tells how many of the parts of a whole have been chosen.

Octagon - a shape with eight sides. A stop sign is an octagon.

Odd number - a number that ends in 1, 3, 5, 7, or 9. Odd numbers are not multiples of two.

P-R

Perimeter - the distance around a figure such as a rectangle or triangle

Pentagon - a shape with five sides

Place value - the position of a number that tells what value it is assigned

Place-value notation - a way of writing numbers to emphasize the place value of each part

Prime number - a number with only one set of factors or only factors of one and itself (One is not considered a prime number.)

Product - the answer to a multiplication problem

Rectangle - a shape with four "square corners" or right angles

Regrouping - moving numbers from one place value to another in order to solve a problem. Also called "carrying" in addition and multiplication and "borrowing" in subtraction.

Right angle - a square corner (90° angle)

Rounding - writing a number as its closest ten, hundred, etc. in order to estimate

S-Z

Square - a rectangle with all four sides the same length.

Triangle - a shape with three sides

Units - the first place value in the decimal system - also, the first three numbers in a large number (starting from the right). The word units can also name measurements. Inches and feet are units of measure.

Master Index for General Math

This index lists the levels at which main topics are presented in the instruction manuals for *Primer* through *Zeta*. For more detail, see the description of each level at www.mathusee.com. (Many of these topics are also reviewed in subsequent student books.)

Addition
    facts .... Primer, Alpha
    multiple digit................................. Beta
Additive inverse .... Epsilon
Angles .... Zeta
Area
    circle .... Epsilon, Zeta
    parallelogram .... Delta
    rectangle .... Primer, Gamma, Delta
    square .... Primer, Gamma, Delta
    trapezoid.................................... Delta
    triangle...................................... Delta
Average .... Delta
Circle
    area................................ Epsilon, Zeta
    circumference.................. Epsilon, Zeta
    recognition .... Primer, Alpha
Circumference .... Epsilon, Zeta
Common factors .... Epsilon
Composite numbers...................... Gamma
Congruent .... Zeta
Counting............................ Primer, Alpha
Decimals
    add and subtract .... Zeta
    change to percent............. Epsilon, Zeta
    divide .... Zeta
    from a fraction .... Epsilon, Zeta
    multiply...................................... Zeta
Divisibilility Rules........................... Epsilon
Division
    facts.. Delta
    multiple digit.............................. Delta
Estimation .... Beta, Gamma, Delta
Expanded notation.................... Delta, Zeta
Exponential notation........................ Zeta
Exponents.. Zeta
Factors .... Gamma, Delta, Epsilon
Fractions
    add and subtract .... Epsilon
    compare .... Epsilon
    divide .... Epsilon
    equivalent............... Gamma, Epsilon

    fractional remainders....... Delta, Epsilon
    mixed numbers ....Epsilon
    multiply....................................Epsilon
    of a number.................... Delta, Epsilon
    of one........................... Delta, Epsilon
    rational numbers .... Zeta
    reduce......................................Epsilon
    to decimals...................... Epsilon, Zeta
    to percents .... Epsilon, Zeta
Geometry
    angles .... Zeta
    area................. Primer, Gamma, Delta,
    .... Epsilon, Zeta
    circumference.................. Epsilon, Zeta
    perimeter .... Beta
    plane .... Zeta
    points, lines, rays .... Zeta
    shape recognition........... Primer, Alpha
Graphs
    bar and line .... Beta
    pie... Zeta
Inequalities .... Beta
Linear measure .... Beta, Epsilon
Mean, median, mode......................... Zeta
Measurement equivalents........Beta, Gamma
Metric measurement .... Zeta
Mixed numbers...............................Epsilon
Money...............................Beta, Gamma
Multiplication
    facts .... Gamma
    multiple digit.......................... Gamma
Multiplicative inverse ....Epsilon
Number recognition.........................Primer
Number writing...............................Primer
Ordinal numbers .... Beta
Parallel lines .... Delta
Parallelogram...................................Delta
Percent
    decimal to percent........... Epsilon, Zeta
    fraction to percent........... Epsilon, Zeta
    greater than 100...................... Zeta
    of a number................................ Zeta

Gamma Index